One Hand Clapping

Unraveling the Mystery of the Human Mind

One Hand Clapping

NIKOLAY KUKUSHKIN

Swift

SWIFT PRESS

First published in Great Britain by Swift Press 2025
Simultaneously published in the United States of America by Prometheus Books,
an imprint of The Globe Pequot Publishing Group, Inc

1 3 5 7 9 8 6 4 2

All rights reserved

Copyright © Nikolay Kukushkin 2025
Artwork by Nikolay Kukushkin

The right of Nikolay Kukushkin to be identified as the Author of this Work
has been asserted in accordance with the Copyright, Designs and
Patents Act 1988.

Printed and bound in Great Britain by CPI Group (UK) Ltd,
Croydon CR0 4YY

A CIP catalogue record for this book is available from the British Library

We make every effort to make sure our products are safe for the purpose
for which they are intended. Our authorised representative in the EU for product
safety is Easy Access System Europe, Mustamäe tee 50,
10621 Tallinn, Estonia gpsr.requests@easproject.com

ISBN: 9781800755000
eISBN: 9781800755017

If two hands come together and make a sound, what is the sound of one hand clapping?

ZEN KOAN

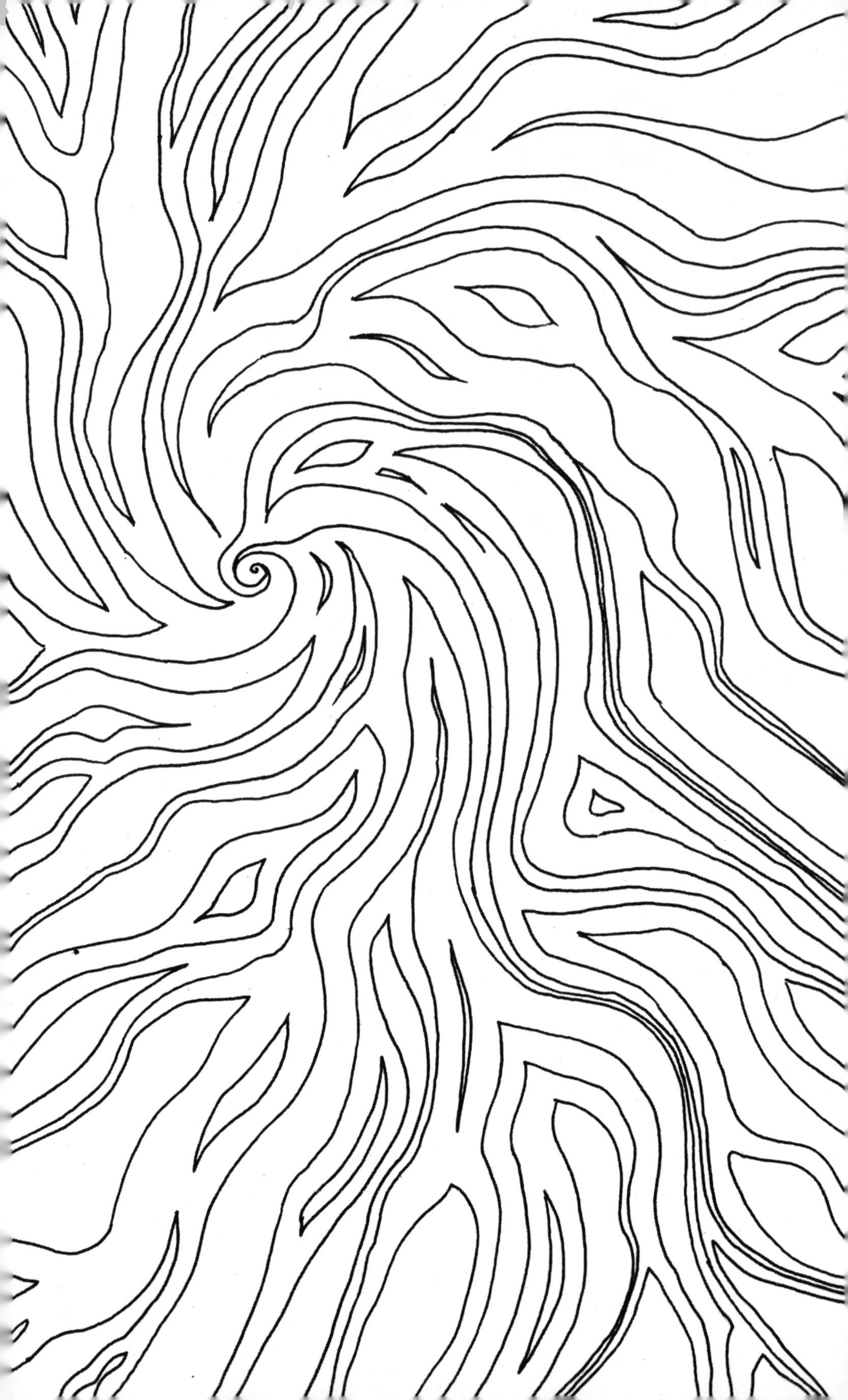

CONTENTS

PROLOGUE — vii

Part I. Where Everything Came From

Chapter 1. In the Beginning Were the Letters — 3
Chapter 2. A Good Idea — 25
Chapter 3. The Birth of Complexity — 39
Chapter 4. When All Else Fails — 57

Part II. Where We Came From

Chapter 5. The Moving Kingdom — 75
Chapter 6. Land! — 97
Chapter 7. When the World Ends — 113
Chapter 8. The Mirror — 143

Part III. Where I Came From

Chapter 9. Animals of Abstraction — 161
Chapter 10. Fire from Within — 181
Chapter 11. The Dark Room — 209
Chapter 12. In the Beginning Was the Word — 229

EPILOGUE — 249
ACKNOWLEDGMENTS — 253
NOTES — 257
INDEX — 277

Prologue

There's a particular type of mental exercise in Zen Buddhism called a koan. It is sometimes described as a riddle without a solution, but that misses the point a bit. True, koans are less about finding the solution and more about suspending yourself in the questioning state of mind, the void of the unresolved. But koans are not meaningless. It's just that the process of getting to the meaning means more than the meaning itself.

If two hands come together and make a sound, what is the sound of one hand clapping?

I first heard this koan from my bohemian intellectual classmates at St. Petersburg State University in Russia. Buddhism was fashionable in a semi-ironic way, postmodern and Westernized. We would swarm the university's famously long central hallway during the breaks between invertebrate zoology and ancient plants and show off our Zen skills to each other with a one-handed clap of fingers against the palm. Oh, the cleverness.

In time I came to see a deeper meaning in the koan. When two hands come together, the sound is what's born out of their contact. Is this not a metaphor for all human experience? In order for me to hear, see, or feel anything, there must be a me, and there must be a non-me. A human and the world that surrounds the human. Experience is what happens when the two come together.

But the koan casts this as an illusion. There are not two things, but only one, and that one thing is making the sound. Neither I, nor you, nor anyone or anything else can be separated from nature as a whole—we are all part of it. To truly understand human experience, implies the koan, we must concede that the boundary between the human and the rest of the world does not truly exist.

I ruminate on the sound of one hand clapping because my scientific work lies on that elusive boundary of me and non-me: I study the miracle by which molecules form memories. In practice, that means I spend a lot of time staring through a microscope at neurons in petri dishes. In many ways, these neurons are no different than those that are wired together in my brain. They connect to each other as they normally would inside my brain and respond to chemicals that, in my brain, would induce memories. When you spend time with these neurons, you soon get a funny feeling—a sort of philosophical Magic Eye illusion. How could it be that what I am studying is at the same time part of what studies it—that is, *me*? Is my mind not run by the same kinds of neurons? Are my memories not stored in the same types of connections? The more I gaze into this junction of me and non-me, the more it seems to melt, running through my fingers like water as I try to grasp it.

Some would say that what puzzles me is the nature of consciousness: the mysterious and possibly unanswerable question of how brains generate first-person experience. But I think the puzzle is deeper than that.

There is a fundamental tension in the way that we think about ourselves and the world around us: *everything in the world is made of the same stuff, and yet it feels somehow special to be me.* This is what puzzles us about consciousness: if I have the same brain as everyone else, why does my *self* feel so special compared to all the others? But it is the same thing that puzzles us when we consider our place in the natural world. If genetically I am not that far from a fruit fly and chemically almost indistinguishable from a mushroom, why does being a human among other species seem so special? If I am made of the same stardust as rocks and oceans, why does it seem so special to be alive?

The levels are different—*self* versus others, human versus nonhuman, alive versus not alive—but the puzzle is the same. From what I can tell, both the Zen koan and scientific research insist that nature knows no true boundaries. But if they are right, then *what is so special about being me?*

Modern science is built upon materialism, the idea that matter is all that really exists. In science classes, the material world is presented as a cold and brutally disinterested space in which lifeless objects collide with one another according to a set of mathematical equations. Even something as lively as the gleam of a butterfly wing can be reduced by science to pigments, synapses, and chemical energies calculated to the second decimal point. Suggest that the gleam is anything but that, and it sounds like you are denying science and probably believe in ghosts or miasmas.

But my book puts forward another way of looking at the world. It emphasizes not the things, but the *ideas*—or what I call the *essences* (from the Greek *eidos*)—inherent in things. There's nothing ghostly or unscientific about that. It is simply a shift in perspective, but it's a shift that allows us to uncover the *meaning* of natural processes. Here's what I mean. You could say: "A muscle in a leg of a human clad in blue and white propelled a spherical object into a network of synthetic strands in a metal frame." Or you could say, "Messi scored a goal for Argentina in the World Cup finals." The first perspective is materialist. We can use it to predict, say, how the trajectory of the ball varies depending on the angle of the kick. But the second perspective is the one to use if we want to understand the meaning of the action. And it is this perspective we will use to resolve our puzzle. To understand what is special about the human experience, we have to see what the human experience *means* from the vantage point of nature as a whole. And to get there, we must start at the very origin of life.

In this book, we journey from life's beginnings forward through the eons until we arrive at the stupendous capacities of our own minds, here, now. Our journey proceeds in stages that are not simply geological eras nor successive branches on the evolutionary tree of life, but are rather stages in the slow crystallization of essences, "nature's ideas." With time, they become progressively more specific, nesting into each other like a set of Russian *matryoshka* dolls: being alive, being animal, being human, being a *self*. As we move forward through these nested boundaries, it becomes clear that the human experience is special not because of any *one* of them, but because of the complete set.

And then, something curious happens. When we reveal the entire *matryoshka*—and thus glimpse the human experience from the vantage point of nature—we see that what makes you and me different from the rest of the world are the very same things that we contribute to its flow, to its unified, eternal river of essences. What we think separates us from nature, in fact, makes us inseparable from it.

I believe this unification of the world, both inner world and outer world, into a single, fluid oneness, is finally what the koan is trying to achieve. Intuition tells us that two hands must come together to make a sound, but in this book, we see that all sound in the world is made by a single hand.

Part I.
Where Everything Came From

Chapter 1.
In the Beginning Were the Letters

Everything happens accidentally.
LEO TOLSTOY *War and Peace*

When I first arrived in the United Kingdom for graduate school, one of the things that baffled me, as a Russian, was that everyone thought vodka was made from potatoes. Eventually I realized that there were all kinds of vodkas including the Polish potato kind, but at the time it seemed barbaric—every Russian knows vodka is made from wheat. I would walk around and ask people what vodka was made from, and when the British invariably said "potatoes," I would rage in the way only a twenty-one-year-old can. The thing was, it all tasted the same. Which starch you use to make the alcohol really doesn't matter all that much. It was the idea that was offensive. The idea that the starch should come from wheat, to me, seemed like an inherent part of what vodka was, regardless of the molecules it was made of. If the idea was gone, the vodka was no longer vodka—although it admittedly still worked for all intents and purposes.

Some might say ideas like this only exist in our minds, not in reality. Visitors from another planet would not be able to distinguish a wheat-based vodka from a potato-based vodka, therefore the two are not *really* distinct substances—so the argument goes.

3

But say the visitors are so advanced that they have access to all human brains and every signal passing through them; they can decode all our languages and read all our books; they can infer the role of wheat in Russian history, and potatoes in Polish history, based on soil chemistry, climatic observations, Chekhov's plays, and the menus of local restaurants. If they can actually take all of this into consideration, they can certainly also see the two vodkas as being meaningfully different. So, in fact, the ideas *do* exist in reality—just not in the material composition of vodkas alone. To get to these ideas, the interplanetary visitors would have to understand a lot more than chemistry.

What we are about to do in this book is similar. We will look for a perspective broad enough to reveal "nature's ideas" about what it means to be a human. Not the ideas that exist in our heads, but the ideas embodied in the world itself, in its flow, its patterns, the causal connections between its elements. Plato called nature's ideas *eidos*, or "essences." This is what I will call them, too, to distinguish them from the ideas of human beings.

Ancients had no issue with nature possessing ideas, rationality, creativity. But modern science classes trained us to reject any attempt to think of nature in rational terms. I once heard a respected professor at a conference boast during her keynote speech that she taught her students to never ask "why" questions. We are taught that things simply exist and asking their "point" is paganism—Zeus and sea-foam, not chemistry and biology.

But I think the ancients had something with all the paganism. To me, it is less silly to imagine the forces of nature in humanlike or divine form than to deny that there's anything rational in nature at all. Whether you use the language of gods and spirits or the language of molecules and mutations, ideas—essences—permeate the world both living and nonliving all the same.

To Build and Destroy

Consider metabolism, one of the key features of any living organism. This glorious system allows us to stuff into our mouths virtually anything and, somehow, without any effort whatsoever, transform it into thoughts and actions. There are two sides to metabolism: anabolism, the buildup of big molecules from small ones, which requires energy, and catabolism, the breakdown of big molecules into small ones, which releases energy. Creation and destruction, two pure essences, like yin and yang. In its specifics, metabolism is extraordinarily complex—hundreds of carefully regulated enzymes constantly interconverting nutrients and energy to ensure that everything is always balanced. But at its core, metabolism boils down to something much simpler—so simple that it is written into the periodic table of chemical elements that hangs in every chemistry classroom in the world. If ancient Taoists knew the periodic table, I'm pretty sure they'd hang it in a temple.

The yin and the yang, creation and destruction, correspond to two atoms of central importance to life on Earth and probably on any other planet: carbon and oxygen.

Atoms, in general, are restless and needy creatures. They are always looking for something they can get from other atoms. As some readers will remember from high school physics, different kinds of atoms (also known as elements, such as carbon, oxygen, or iron) have a designated number of electrons, which hover like a cloud around the protons and neutrons sitting in the nucleus. Whatever their allotted number of electrons, atoms never seem to get it right and are always anxious to make some changes. The entirety of chemistry is some form of atomic hustle with electrons. Some atoms want to steal an electron from someone else. Others want to offload an electron surplus. Others yet look to form an electron-sharing partnership with another atom—a molecule. This—assembly into molecules—is one of the ways by which atoms control their anxiety.

Molecules of *life*—organic molecules—are distinguished by their large size. They are made up not of two or three atoms but tens, hundreds, even thousands of atoms with their electron clouds arranged into complicated three-dimensional structures. All of that is possible thanks to the properties of carbon.

In the broadest terms, life really is made of carbon, with a few other things stuck to it. Why carbon? It's a uniquely cooperative element. A carbon atom can form four separate chemical bonds with other atoms, which is more than most elements. What's equally important is that those partner atoms can also be carbons. Very few elements so readily bond with their own kind. The result is an infinite array of complex molecules with branched multicarbon chains and polygonal multicarbon rings of virtually unlimited shape and size.

You could say that as an element, carbon is practical, enterprising, and constructive. In Russian, the names of elements come with a preassigned gender,

C
(carbon)

O
(oxygen)

H
(hydrogen)

H₂O
(water)

O₂
(molecular oxygen)

CO₂
(carbon dioxide)

which certainly charges the air with gender stereotypes, but also adds an inescapable humanizing element. Platinum is female, iron is neuter, carbon is male. I have always thought of carbon as a sturdy middle-aged man, maybe a coal miner. In his dealings with other elements, carbon is reasonable and collaborative. He is not looking to steal anyone's electrons, only to unify electron clouds into a communal whole, building impressive molecules of increasing scale. He shares equal rights with other carbons within their complex rings and branching chains. Carbon is even nice to hydrogen, pulling only slightly on his laughable single electron. (Hydrogen, the simplest element, consists of a single proton and a single electron—small and almost powerless in the chemical sense.)

Oxygen is the yang to carbon's yin. He—oxygen is also male in Russian*—is destructive, merciless, and ferocious. He rips apart anything in his way. The force with which oxygen pulls electrons toward itself is second only to fluorine,[1] itself a much more exotic character. As oxygen rams into molecules, every complex arrangement is destroyed, every chemical bond broken, atoms separated, their electron clouds sucked into oxygen's insatiable orbit. The energy stored therein is liberated with spectacular gusto, sometimes in the form of light and heat—that's called combustion.

Take the essence of carbon, take the essence of oxygen, put them together, and you get the essence of metabolism, one of the most magical features of life on Earth. Carbon builds, oxygen destroys. This describes their properties anywhere in the universe, but it also describes how our bodies work at the most basic level: we make large molecules out of carbon and then break them down using oxygen.

There is no dichotomy more *elemental* to life on Earth than this one—whether you use the word in the chemical or philosophical sense. Everything else is built upon these two. Carbon's cooperative nature means a limitless chemical palette from which life can draw its form. Oxygen's propensity to break and grab means a constant turnover of old into new, and by extension, the impossible into the possible. Carbon represents the enterprising spirit that sets the tone for all the delightful innovations of life. Everything in living nature that grows, or branches, or builds up, anything that challenges the norm, anything that extends

---

*As are most elements. Besides platinum, the only females are sulfur, copper, mercury, and antimony, whereas gold, silver, tin, and iron are neuter.

(part of a complex organic molecule)

combustion

beyond the previously possible owes its existence to the atom of carbon and follows in its pioneering footsteps. But oxygen is equally vital to nature as we know it, just as death is part of life. Oxygen represents a morbid but unshakeable principle: for something to be created, something else must be destroyed. For anything big, anything complex, anything beautiful that exists in nature, there is a price to be paid, a resource to be used up, energy to be spent. Oxygen provides that energy. For carbon to build, oxygen must ravage. Oxygen is not just nature's bully; he is a chemical Shiva—bringing rebirth through destruction. The two elements, in large part, define what it means to be alive.

The Circle and the Dead End

Another signature feature of life on Earth is that it obeys the so-called central dogma, the closest biology comes to a physical law. This dogma captures the most puzzling feature of life's molecular makeup. Francis Crick, co-discoverer of the DNA double helix, who first formulated the central dogma in 1957, put it best: "once 'information' has passed into protein, it cannot get out again."

Let's unpack. The "information" to which Crick refers is genetic information or, simply, genes: the heritable sequence of "letters" of DNA.* This information gets passed from one generation to the next by copying said DNA. But then, in each generation, the same information is also "passed into protein." Proteins are life's principal molecular machines, whimsical nanorobots with which everything in a living organism is done. Genes are basically blueprints for these proteins: the sequence of DNA letters is a code that is used to assemble them. This is what "information passing into protein" means. And here's the dogma: once the code is in, it can't get out. You can't extract the blueprint from the protein and make another protein based on that.

Note that the statement that Crick chose to encapsulate the central dogma is not about what DNA can do (pass its contents from generation to generation and serve as a blueprint for proteins), but about what proteins cannot do: *proteins cannot copy themselves.* They are an informational dead end. Every protein eventually gets destroyed—either because it is no longer needed or

---

*It is curious, but not surprising, that Crick chose to put the word *information* into quotation marks: back in 1957, at the dawn of modern genetics, this computer-science language was still very new, and applying it to genes—to *life*—seemed like science fiction. Today no one would flinch.

20 amino acids

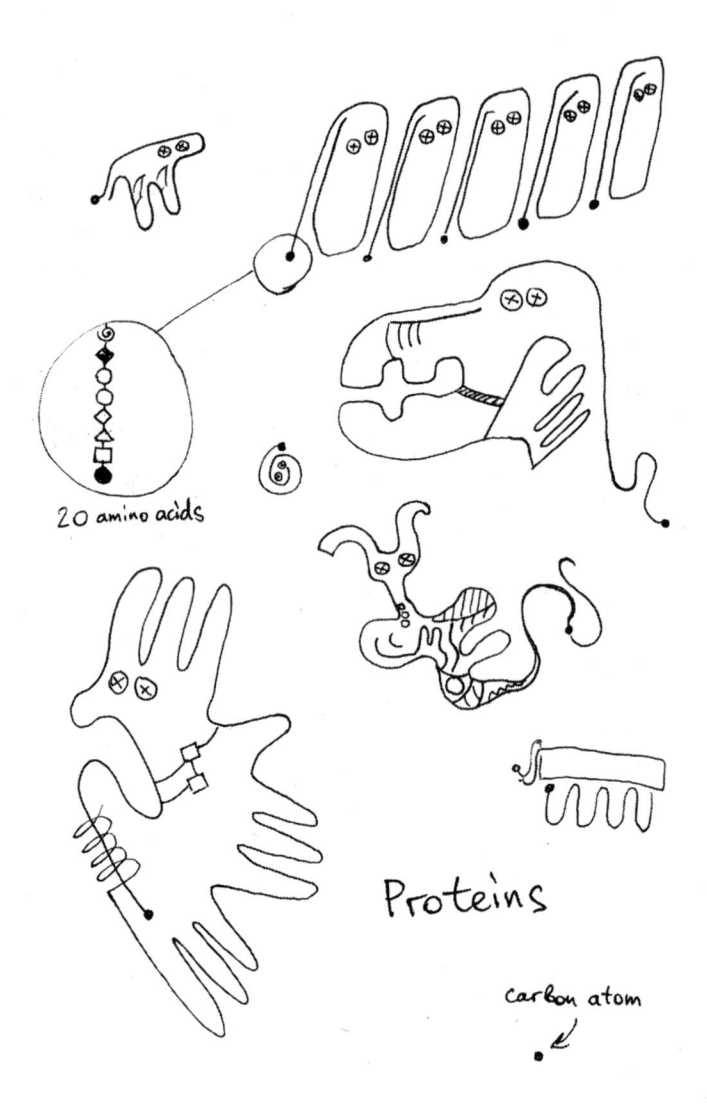

Proteins

carbon atom

10 One Hand Clapping

simply due to wear and tear. When that happens and a protein falls apart, the information inside of it dies, and a new protein must be made again using a DNA blueprint. A protein by itself cannot carry a gene onward into the future. The reason this is so significant—and puzzling—is because in a living organism, *proteins do nearly everything else.*

A protein is not just one specific substance; it is, rather, a type of molecule. Proteins are all the different chains that can be made from the same set of much smaller molecules called amino acids, which are strung together in sequence, like beads. There are a total of twenty different amino acids that occur in proteins (think twenty colors of beads). Amino acids are all simple molecules but quite different from each other in their chemical properties. A typical protein is made of several hundred of them—assembled in a particular sequence, it contorts into an intricate three-dimensional shape spattered with chemical groups operating as the gears and cogs of a machine. Humans have roughly twenty to twenty-five thousand different proteins in total,[2] and there is a gene for each one, a corresponding region of DNA that serves as an instruction for assembling this specific sequence of amino acids. Each cell decides for itself which subset of these twenty thousand to produce, in what quantities, and at what time according to its needs.

These diverse proteins rule the living organism. Like workers of different professions, they do anything and everything there is to be done. We digest food using proteins, breathe in oxygen using proteins, and move using proteins. Proteins identify viruses, proteins synthesize the cell membrane, and when long-term memories are formed, proteins in the hippocampus are using proteins to send protein signals to other proteins in the cerebral cortex.

And in the most intriguing twist, proteins are also in charge of copying DNA—that very thing they cannot do for themselves.

Like proteins, DNA is a chain of molecules, in this case called nucleotides, strung together in a sequence. DNA chains are larger and clunkier than protein chains, and although they contain instructions for making proteins, they cannot do very many things on their own. Most of the time, DNA just floats there, while proteins climb all over it, making stuff happen—reading the sequence, repairing the sequence, copying the sequence. Without these proteins, DNA is almost helpless and certainly could not organize its own replication. So proteins are responsible for propagating DNA, which in turn holds the key to the existence of proteins. DNA needs proteins so that it gets copied, but proteins need DNA so they are re-created in each generation. That's the ultimate chicken-and-egg.

Why this bizarre arrangement? If proteins are such universally capable molecules, if they are so much *better* as molecular devices than DNA, why can't everything, including inheritance, just run on proteins? This, by the way, was the predominant theory until the 1950s, when the role of DNA was definitively proven. The reason for this is a key feature that DNA possesses and proteins lack. It is called complementarity, and it is the axis on which the circle of life spins.

DNA has only four different nucleotides, which is a whole lot fewer building blocks than in proteins. These blocks, nucleotides, are also not as diverse as different amino acids are—chemically, they are more or less similar molecules. What they have instead is this key property, complementarity, also known as base pairing. The four DNA nucleotides, known as A (adenine), T (thymine), C (cytosine), and G (guanine), are organized into pairs that stick to each other: A sticks to T; C sticks to G. This doesn't seem like much, but it means everything.

Because each nucleotide has a complementary counterpart, any sequence of nucleotides also has a complementary version—for example, ATTCG is complementary to TAAGC, like a positive and a negative. If you have one sequence, you can create the other using the first as a template, and vice versa. A typical DNA molecule carries both a "positive" strand and a "negative" strand—two complementary chains stuck to each other and wound into a double helix. Unwind the helix—and you have two complementary chains. A special DNA-weaving protein, or DNA polymerase, comes along and rebuilds the missing strands—a positive to a negative, and a negative to a positive, according to the same simple rule: A to the T, C to the G. Voilà—you end up with two identical double helixes. This is how DNA replicates, and it is only possible thanks to complementarity.

It is humbling to think that the continuity of generations, sustained for billions of years, connecting each living creature to our common ancestors and the very origin of life on Earth, hinges on four small molecules, the letters of the genetic alphabet, sticking to each other in pairs. In a way, the complementary chains of DNA represent the very essence of life. Think about it: it is only in biology that multiplication and division are the same thing, thanks to DNA. To multiply, living beings divide. That doesn't happen when, say, snowflakes multiply in the air or when dirty dishes multiply in the sink. But new living organisms always in some way bud off already existing ones, ultimately—because DNA is copied by splitting the original in two.

Proteins don't have anything comparable. Amino acids don't come in complementary pairs, so there is no way to make a replica of an already existing protein: information contained within them "cannot get out," per Crick's formulation of the central dogma.

In other words, *proteins don't have access to eternity.* That—eternity—is why they need DNA, whose paired nucleotides provide just that.

On the other hand, DNA without proteins is inert and lifeless. It is only thanks to their extraordinary abilities that DNA can take advantage of its complementarity, replicate, and impose its genetic will on the living organism. So what DNA needs proteins for is their nanorobot-like chemical versatility—which ultimately boils down to their tool kit of diverse amino acids.

So, in their very chemical nature, nucleotides and DNA embody continuity, whereas amino acids and proteins embody functionality. These essences can be separated, but one cannot exist without the other, and life as we know it cannot exist without either of the two.

One might think of DNA and proteins as equal partners in the industry of life. Actually, the relationship between these two great molecules of nature, and between the essences they embody, is more complex. DNA and proteins are not equals. A random change in DNA, as in a mutation, means a change in all proteins encoded in it, a change that could persist forever. But a random defect in a protein is as short-lived as the protein itself. Once the defective molecule falls apart, it does not affect DNA, or future generations, which continue to produce the same protein without any alteration. At the end of the day, it is DNA that controls proteins, not vice versa.

That's really quite tragic. Proteins, these marvelous molecular machines capable of almost anything except self-replication, are forced, because of this deficiency, to labor for the benefit of the genes, controlled by their needs, subject to their whims. *That which reproduces holds the power.* Later in this book, we see this rule play out again and again: in the relationship between worker ants and their queen, in the relationship between the body and its sex cells, even in the relationship between individual experience and culture. Here, in the mutual arrangement of a few atoms in amino acids and nucleotides, the same essence is embodied in its purest, primordial form.

The World before the Dogma

The central dogma—the rule that genes "flow" from DNA to proteins and not vice versa—is usually represented in biology classrooms with a flowchart containing an extra level in the middle: DNA, *to RNA*, to proteins. RNA is DNA's cousin, a similar molecule made out of slightly different nucleotides—the "NA" in both acronyms stands for "nucleic acid," and the first letters—"R" for "ribo-," "D" for "deoxyribo-"—refer to these small differences in the chemistry of RNA and DNA's building blocks. RNA sits in the middle of the central dogma flowchart for no obvious reason. DNA is for inheritance; proteins do the jobs. You could imagine DNA directly converting into proteins. But that's not how it happens. In reality, DNA is converted into proteins through the medium of RNA. First, a region of DNA must be *transcribed* (essentially, printed out) into its RNA equivalent. Then the printout must be *translated*—the sequence of RNA converted into a sequence of amino acids.

This two-step conversion, in itself, does not make a lot of sense. It seems like one of those things that you are just required to accept in a science class without asking "why." That's just how it is! But I can't think of a better place to ask "why" and to look for a reason RNA exists. Because the answer explains not just how molecules work—it tells us how our story on this planet begins. The thing is, RNA might have been the original form of life. The reason it's still there is because everything else grew around it.

Let's back up a little. To get from a gene to a protein, first, you have to transcribe the gene into an equivalent string of RNA. This is straightforward because RNA and DNA are so chemically similar. Just like you can copy DNA using complementarity (A to the T, C to the G), you can transcribe it into RNA in the same way (except RNA has a slightly different nucleotide, U instead of T, though the two are functionally equivalent). All you have to do is find a gene you are interested in on the long meandering helix of DNA, unwind that part of the helix, and duplicate one of the strands using RNA nucleotides, like a photocopy of a page in a book. The photocopy—RNA—then peels off the book—DNA—and prepares to be converted into protein.

This next step, however, is much more complicated. Proteins and nucleic acids are totally different. There's nothing like "A to the T, C to the G" to guide the assembly of one molecule based on another. You have to convert a sequence of nucleotides into a sequence of completely unrelated molecules—

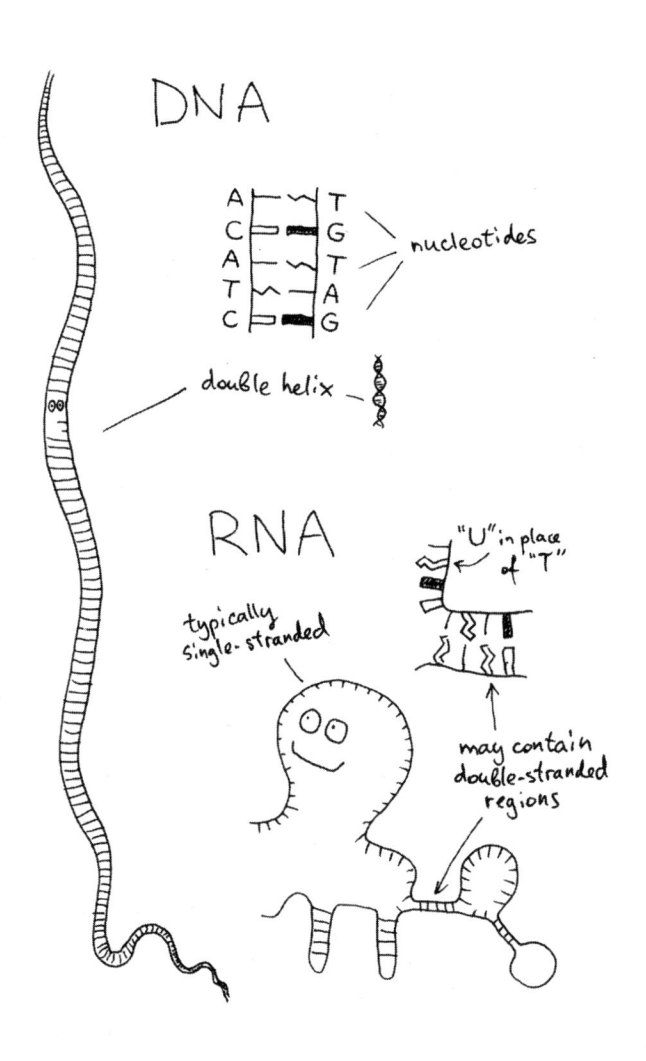

amino acids, which make up the proteins. It is like translating one language into another, and it requires some high-grade molecular trickery. It happens at an all-important cellular factory called the ribosome, a large, oddly shaped molecule that takes in the RNA printout of a gene and converts it, letter by letter, into the amino acid sequence of a protein.

This is where it gets especially interesting. Here we have the ribosome, a critical element of a living organism, a wondrous protein-making factory that interconverts two molecular languages. Virtually all molecular machines in nature are proteins: proteins do all the jobs, including even copying DNA, which can't achieve replication on its own. You would think that the job of *creating* proteins would also belong to proteins. And yet, surprisingly, this is not so. Instead, the ribosome employs RNA—hence the "ribo" in its name.

This is extremely unusual. Generally speaking, nucleic acids are not good at doing jobs—any jobs. DNA is especially inert and would never consider anything as flamboyant as being part of a protein-making machine. RNA is not that much better—it is also made of nucleotides, and like any nucleotide chain, it is big, awkward, and nowhere near as versatile as a protein. Nucleic acids are almost always used as information carriers, not as components of functional devices, which, in turn, are almost always protein based.

But there are exceptions to the rule. Sometimes RNA acts surprisingly like a protein: it doesn't just carry a code in its sequence, but actually *does something* as a molecule, a clumsy visitor from the high-society nucleic acids, getting its hands dirty with protein-esque manual labor. Some RNA molecules even look like proteins: instead of long formless strands of the DNA kind, they fold into compact three-dimensional shapes, very much in the protein fashion. The ribosome is a perfect example—the most notable enterprise organized by such protein-like RNAs. This protein-making factory is actually a conglomerate of different molecules, but RNA runs the show: it does the most critical jobs of selecting the appropriate amino acids and connecting them to one another during the protein assembly process.

What is going on with RNA, this protein wannabe in the nucleic acid family? You can imagine the central dogma flowchart with DNA and proteins alone, and yet RNA stubbornly intervenes and in fact holds the key to the entire "flow of information into protein." Why is it there at all? The reason, say biologists who favor the "RNA world" theory, is that RNA was the original, most ancient life-form that had ever existed on our planet and, simultaneously, the prototype of both proteins and DNA, which evolved later as more specialized extensions of RNA's abilities. In other words, RNA is a relic of the origin of life, much like the cosmic microwave background is a relic of the Big Bang.

At first glance, RNA compares unfavorably with both DNA and proteins. It's not great for storing genes in the long run because it is less stable than DNA. In today's world, only some viruses with very simple genomes are able to store their genes in the medium of RNA, and those viruses (COVID and flu, to name a couple) mutate and evolve a lot faster than viruses that opt for DNA (for example, rabies—this is why getting vaccinated once is good for life). RNA is also not as good as proteins at doing jobs because its nucleotides are chemically inferior to proteins' amino acids. So DNA is a better archive, and proteins are better machines than RNA.

But what is profoundly unique about RNA is that it can be an archive and a machine *at the same time*. It can embody the essences of DNA and proteins—continuity and functionality—within a single physical unit.

It is for this reason that evolutionary biologists love RNA as much as they do. Since RNA can both *do things* and *be replicated*, the easiest way to imagine the advent of the "central dogma world"—today's world with its codependent trio of DNA, RNA, and proteins—is to start with self-sufficient RNA that replicates itself.

Maybe it exists alone, multiplying only its own sequence, and its many copies gradually diversify. Maybe it altruistically replicates every random RNA it can find. In either case, over time, many different RNAs are created and replicated together. They take on a variety of molecular jobs that aid their collective reproduction. Then, eventually, comes the greatest milestone in the history of this "RNA world": the invention of proteins.[3] The advent of the ribosome—in its original form, a complex and folded RNA machine—enables RNAs themselves to transform their sequences—today known as genes—into protein sequences, producing an unlimited number of amino acid–based nanorobots. This invention opens a new world of possibilities for RNAs to create novel functions and optimize already existing ones. Almost all work is then relegated from RNA to proteins, save for a few rare instances (such as in the ribosome itself). A great variety of new jobs is created: proteins learn to replicate RNA, mint nucleotides, harvest and store energy, and eventually to create cell membranes and all the other essential components of a living organism. Finally, proteins create a new, superior, highly stable archive for storing genes: double-stranded DNA. The "central dogma world" as we know it is complete.

All in all, if we accept that RNA world is how things started, given a billion years or so plus some imagination, you can probably get from there to everything else.

But how do you get to this presumed starting point—a self-sufficient, self-replicating RNA? Is it even possible? The answer appears to be yes. Scientists have been able to artificially create an RNA system that can self-replicate indefinitely without any help from proteins.[4] The nuance is that this system is not just a single self-sufficient RNA molecule making copies of itself, but rather several molecules that achieve their replication collectively. To simplify somewhat, molecule A replicates molecule B, B replicates C, and C replicates A, so no single molecule is sufficient, but the combination is locked in a loop that replicates all its members.

This scenario actually seems even more realistic vis-à-vis the origin of life. In nature, almost everything begins with an accident, but almost every accident ends in nothing. For there to be anything useful, there must have also been many RNAs doing many random things that never took off. Rather than imagining that one day among them a single Promethean benefactor started replicating everybody else, we are invited to imagine a soup of random and diverse RNA chains minding their business and going about their own agenda set by their randomly assembled sequences, until one day this soup finds itself interconnected through mutual replication. One day just the right combination of biochemical activities falls into place and forms something like the self-replicating collective the scientists created: A happens to make more B, B to make more C, and C to make more A—and there could have been thousands more RNAs involved in this collective cycle. Call this the "loop in the soup" model. Once the replication loop is formed, the circle of life starts spinning and continues to this day.

But just because this is achievable deliberately in the lab does not necessarily mean it is achievable spontaneously in real life. What could this primordial RNA brewery possibly be, in a real, physical sense? In search of the answer, most experts look to the deep sea.

The moment life began

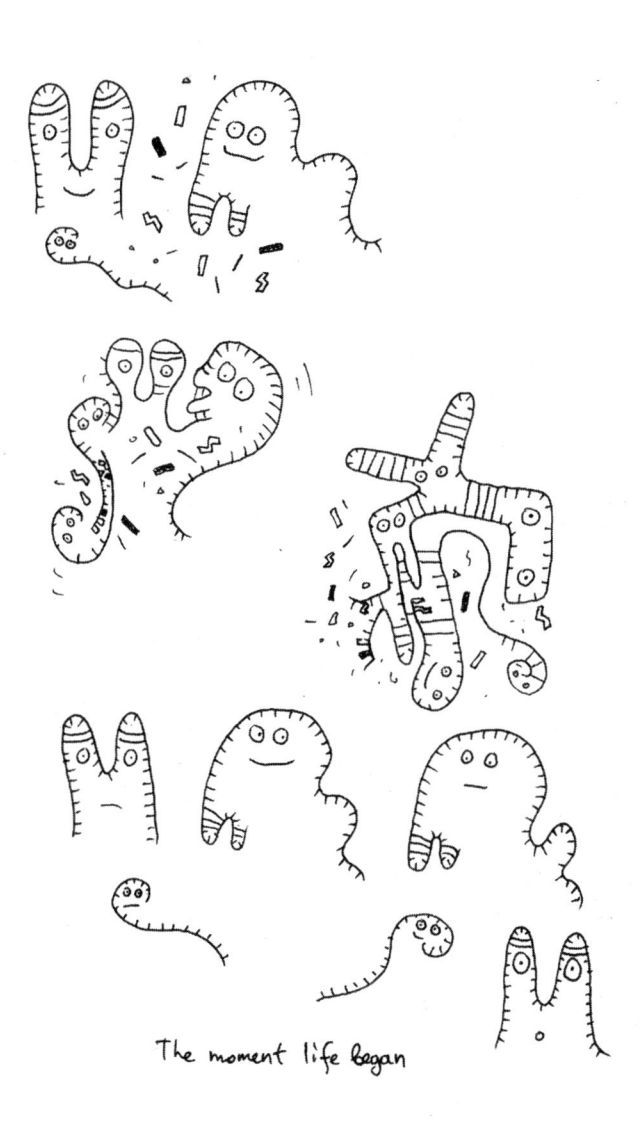

In the Beginning Were the Letters **19**

Warm Little Vent

To create a soup of RNA capable of "going live," you need to solve several problems. First you need to spontaneously create nucleotides, the "letters" from which chains of RNA could be assembled. On modern planet Earth, any spontaneously created complex molecule would either get immediately broken down by oxygen or eaten by some microorganism, but neither was present on the primordial planet, so there was a bit more room for spontaneity. The best-known evidence that *something* can be spontaneously created is the famous Miller-Urey experiment. In 1953 chemists Stanley Miller and Harold Urey demonstrated that if you take some simple chemicals, seal them in an oxygen-free tube, and proceed to heat and zap them with electricity (emulating lightning on ancient Earth), eventually a whole array of complex organic compounds will be produced. Miller and Urey didn't quite get to nucleotides, and for a while the spontaneous generation of those particular molecules seemed implausible. But based on recent research, at least the bulk of a nucleotide can be spontaneously assembled from simpler molecules under conditions resembling what we know of ancient Earth, so it's beginning to look more and more possible.[5]

Next, there must be enough raw materials to mold into these nucleotides and ultimately RNA. Beyond carbon, oxygen, and hydrogen, you need phosphorus and nitrogen, and ideally a variety of other microelements that might facilitate chemical transformations. Modern organisms rely on a good part of the periodic table, including exotic elements like copper and manganese required in small quantities for specific chemical tricks they are good at.

There must also be some source of energy—a force that could turn simple molecules into complex ones. Miller and Urey imagined lightning that would strike "warm little ponds" (a phrase originally uttered by Charles Darwin himself). Ponds are attractive for an additional reason: spatial constraints. Life as we know it is all based on water, so it must have started in the water. But it could not have started in the open ocean, because all participating molecules must be kept in one place, as they are in a modern organism by our cell membranes, blood vessels, and gut linings.

So you need water, chemicals, and energy in a tight space. A little pond struck by lightning fits. Another place that fits even better is a deep-sea hydrothermal vent.

Although there are many different opinions when it comes to life's birthplace, as of this book's writing, the most popular theories are associated with these mid-ocean geysers that spew hot water and minerals from deep in the Earth's crust.[6] Vents seem to embody several ideas foundational to today's life-forms, and for that reason many questions about the origin of life can be answered by finding just the right type of vent. It is almost as if a hydrothermal vent were a prototype of a living being.

First, hydrothermal vents are chemically rich. Vents are tapping into the crust of the Earth, so the water they spew contains a variety of substances and chemical elements. These often get deposited around the water source, producing an array of impressive structures of different shape, size, color, and chemical composition, depending on the water's mineral content. This allows for a wide variety of chemical reactions that would otherwise be considered unlikely, which, from a chemist's point of view, is also a trademark feature of life.

Second, hydrothermal vents build up around a source of energy, much like living organisms. Their discharge contains energy in both chemical form, such as hydrogen gas, and in the form of heat. The temperature differential—between the heat inside the vent and the cold water of the surrounding ocean—might itself have played a role in the origin of life, since it provides a way to constantly stir the hypothetical soup of primordial RNA, much like today circulation stirs the hormones, nutrients, and oxygen in our bodies.

Finally, the hydrothermal vents offer even better spatial constraints than "warm little ponds": the columns that build up around the geysers are porous, and the typical size of the pore is remarkably similar to the size of a modern cell, such as those in our multicellular organisms. These tiny mineral bubbles may have been the original testing labs for the development of life. Even more compellingly, the surface inside the pores is adsorbent, which makes it more likely that hypothetical nucleotides would assemble into hypothetical RNA chains—it would be easier for the "letters" to find each other and combine into sequences while clinging to the pore's surface.[7]

To picture a representative location, think of the Lost City, a mid-Atlantic "hydrothermal field" discovered in 2000.[8] Although the "city" itself is relatively recent (not even mountain ranges have existed for *billions* of years), some experts argue that sources of this type most resemble the ancient cradle of life. The Lost City is an eternally dark, five-hundred-square-meter area at the bottom of the ocean studded with numerous hydrothermally derived structures: pipes, chimneys, and cathedrals as tall as a twenty-story building. Some of them are hot and active, others are cool and dormant, and the entire field is populated by scores of extraordinary creatures, most notably a whole jungle of unusual microorganisms. Although we will never know, at the very least it seems possible that our history begins in a similar place. Are there new lifeforms brewing in the depths of Lost City as we speak? The rules have changed in the past four billion years: the planet is now taken over by ever-hungry living organisms and saturated with that ever-destructive gas oxygen, both of which make it seemingly impossible for entirely new forms of life to emerge. But perhaps we are underestimating the power of chance.

The Junkyard Tornado

For almost all of human history, at all stages of humankind's cultural and scientific development, the existence of life was seen as a miracle—something beyond the explanatory power of common sense. Until the nineteenth century, even secular scientists believed that life must possess some special force that fundamentally distinguishes it from nonlife—today we call this rejected school of thought "vitalism." Modern science says that the gap between life and nonlife can be breached—this transition is called abiogenesis.

I think the reason abiogenesis is such a contentious topic outside of scientific circles—it's been likened to a tornado assembling a Boeing in a junkyard—is that many people feel that it erases the distinction between life and nonlife and, in doing so, takes away something that makes human existence special. Surely the miracle of life is not just all those dull atoms and molecules!

But if we recognize that atoms and molecules can *mean* things—that they can represent nature's essences—then the continuity of life and nonlife ceases to insult the dignity of life in any way. Instead, it connects each of our individual lives to profound, elemental forces that define what we are in the broadest of terms. Look at the properties of carbon and oxygen, and see the contours of metabolism, the cycle of matter and energy. Look at nucleotides and amino acids, and see the foundation for the central dogma, the life cycle of continuity and functionality. Look at a hydrothermal vent, and see an outline of a living organism. Human existence can be special and life miraculous not despite these connections but because of them. What they reveal is that ideas don't exist only in human heads but are inherent in all things, even the simplest ones and even the nonliving ones. The essences of life stem from the essences of nonlife, and do so in a continuous, unbroken flow that begins long before the origin of living nature.

As for how life came to exist, the theories might change in the future. But there's one thing about the origin of life that requires no proof other than, perhaps, a walk in the forest: life *happened*. At that, I gasp in awe.

Chapter 2.
A Good Idea

> Listen!
> If stars are lit,
> it means—there is someone who needs it?
>
> VLADIMIR MAYAKOVSKI

The lab where I did my doctoral research was located within a three-minute walk from the site of the greatest clash between evolution and religion in scientific history: the Oxford evolution debate of 1860. This battle between proponents of the Darwin's theory of evolution and Anglican defenders of the Bible took place at the Oxford University Museum of Natural History, or University Museum, as it was then known, on the day of its opening, June 30, 1860, as if it were specifically built for the epic showdown. The Oxford debate is often considered the historical moment when humankind pivoted from the theory of intelligent design to the theory of evolution.

The battle, according to Darwin's friend, Joseph Dalton Hooker, who attended the debate, "waxed hot"—blood boiled and ladies fainted. What so animated the debaters and spectators alike—there's no precise record, but there were maybe five hundred or even a thousand of them—was the recent publication of Charles Darwin's magnum opus, *The Origin of Species*.* This wasn't one of those cases in which a great idea goes unnoticed for half a century.

---

*Origin's full name is *On the Origin of Species by Means of Natural Selection, or the Preservation of Favored Races in the Struggle for Life*. People had more time back then.

Darwin's theory caused a furor the moment it went into print. Everyone was talking about it, and passions flared. On stage at the museum, the conservatives, headlined by the Oxford bishop Samuel Wilberforce, mocked the evolutionists' supposed ape ancestry. The evolutionists (led by the London biologist and self-proclaimed "Darwin's Bulldog," Thomas Huxley, who remains etched in history not in least part thanks to this catchy canine alias) fired back that evolutionary theory is plain logic. Neither spared any vitriol.[1] But the most memorable moment of the debate may have been offered by "a grey haired Roman nosed elderly gentleman," who, according to Hooker's account, stood in the center of the audience and shook an "immense Bible" over his head, castigating Darwin for his ungodly lies, until he was booed away.[2] This was none other than Robert Fitzroy—by then, rear admiral, but some thirty years earlier, a twenty-six-year-old captain of the Royal Navy, freshly appointed commander of HMS *Beagle*.

The *Beagle* had a dark past—its previous captain shot himself in his cabin—and Fitzroy wanted a companion to keep him sane (his own uncle, a notable political figure, also had committed suicide by slitting his throat with a pocketknife, which must have added to Fitzroy's anxiety). Charles Darwin, a twenty-two-year-old geology student, was not his first choice—Fitzroy particularly disliked the shape of his nose—but eventually he took Darwin on board the *Beagle*, setting off the chain of events that would give rise to the modern theory of evolution. At the time, however, Fitzroy, a devout Christian, had no way of knowing of the heretical horizons to which the voyage would lead. The two young men got along. Fitzroy, four years Darwin's senior, assumed the role of a kind of nautical Virgil to Darwin's Dante—a wise guide leading his apprentice through ocean and jungle. It took almost three decades for the insights that Darwin gained during the voyage of the *Beagle* to mature into *The Origin of Species* and culminate in their triumph at the Oxford debate. This must have been a crushing blow to Fitzroy, who surely regretted showing the world to the eventual evolutionary messiah. Five years after the debate, Fitzroy put a razor to his own throat, following in the footsteps of his uncle, as he so feared.

It's hard not to see his end as the ultimate defeat of faith in the face of dispassionate, godless Darwinism. After all, the theory of evolution contends that our fate is inseparable from the fates of our ancestors. Darwin has won. God has lost.

This is the story we have been taught. Humans, we are told, used to believe in a passionate, creative God until Charles Darwin came along and replaced all that passion with a cold, disinterested space of random events that auto-selects the winners and the losers. Darwin put nature on autopilot. In school, we are taught that evolution—the replacement offered for divine inspiration—has no goals. It stumbles blindly into the future with no idea of what might stick. Giraffes have long necks not because they had prudently exercised them in previous generations (as some early biologists believed), but because neck length was initially random, and then short-necked giraffes all died out, leaving behind only the long-necked ones. Ever since the Oxford evolution debate, teachers of science have discouraged their students from thinking of nature as a rational design. Darwin, we understand, disproved God.

But to see post-Darwinian nature in this way—as a dull, soulless automaton—is to miss the point. In fact, you could say that Darwin did the opposite of what we think he did. By exposing to the world the creative power of natural selection—an omnipresent force of innovation that drives the evolution of species—Darwin did not so much disprove God as he inadvertently injected God into every aspect of the physical world.

The Collector of Finches

Newton had an apple fall on his head, Archimedes yelled "eureka!" while taking a bath, and "Pythagoras's pants" have, according to a Russian rhyme, "on all sides—the same expanse," referring to an illustration of the Pythagorean theorem that resembles trousers (it doesn't make total sense in Russian, either—the "sides" of the "pants" do not, in fact, have the same expanse). No such middle-school comedy sketch exists for Darwin, probably because it took him twenty-plus years to eke out a theory after the *Beagle*'s return, which was not terribly conducive to cinematic effect. But if we had to choose, the closest thing Darwin has to Pythagoras's pants is the Galapagos finches.

The Galapagos Islands are a volcanic archipelago in the Pacific Ocean west of Ecuador. These islands appeared recently in geological time and have never been in contact with the continent, so plants and animals have never freely moved between them and the South American mainland. Galapagos does, nevertheless, have plenty of plants and animals—occasional castaways that manage to pass the distance from the continent to one of the islands in the archipelago.

Darwin took great interest in the flora and fauna of Galapagos. In his era, that meant ripping out every plant he saw (for drying), as well as shooting copious birds (for preserving their skins). Based on letters, Darwin seems to have had a blast collecting these bird skins and enthusiastically shot up to ten birds a day, amassing a collection of almost comical proportions. One type of bird that Darwin bagged in great numbers was the finch, a small bird about the size of a sparrow.

There seemed to have been lots of finches on Galapagos. More than other birds. Even more curiously, birds nesting on different islands had small but consistent differences, especially in their beak shape. Some finches seemed to have beaks that were better suited to catching insects, others to a vegetarian diet. One beak, for example, might be sturdy and useful for cracking nuts, another, thin and effective for pecking bugs from cactuses. The differences were especially pronounced when birds with several types of beaks coexisted on a single island.

According to an apocryphal but convenient legend, this observation caused an argument between Darwin and Fitzroy. Fitzroy is supposed to have pressed for independent "centers of creation" on various islands, targeted foci of intelligent design adapted for each island's conditions. On an island with more cactuses, God placed a cactus-feeding bird, and on the island with more nuts, a nut-eating bird. Darwin, though, could not shake off his doubt: was this really part of God's original plan? What if the plan had nothing to do with it?

Galapagos finches have beaks adapted to their food source

What if all these different finches became like this gradually, *descended with modification* from a common ancestor? (Darwin preferred the phrase "descent with modification"—the word "evolution" is only mentioned once in the entire *Origin of Species*, and that in the final paragraph.) What if the different finches were not placed on different islands by divine order, but they adapted to local conditions *themselves*?

There it is, the birth of a great theory!* My favorite way to put Darwinism in a nutshell is this:

variation + inheritance + selection = evolution

First, there must be a *variety* of finches with somewhat different beaks to start with. And indeed, no two finches are exactly alike, as is always the case in nature. Second, there must be *inheritance*: a finch with a large beak

---

*Unfortunately, the truth is less cinematic: Fitzroy never really proposed any "centers of creation," nor did Darwin fully process his ideas about natural selection until much later, and the different species of finches were only identified by a more knowledgeable ornithologist upon *Beagle*'s return.

A Good Idea **29**

must produce large-beaked offspring and a finch with a small beak, a small-beaked one. That is also the case for finch beaks and for any other hereditary trait. Third, there must be *selection*, a preference for one beak or another in a given environment. And in fact large beaks clearly have an advantage on the nut-rich island, whereas birds with slender beaks can gather more food on the cactus-rich island.

But see, says Darwin: if you put one, two, and three together, you *must* get evolution. It is logically inevitable! Better adapted birds multiply faster, and so gradually beaks evolve to suit their environments. *And so does everything else in nature.*

If life is a car and evolution is its motion, then variation is the car's fuel. Without random chance constantly supplying a variety of new possibilities, there's never anything new, so evolution can't move forward—there's no fuel.

Inheritance is the wheels of the car. It's what physically moves life forward, connecting generation to generation, preserving the creations of chance, re-creating them in new physical form again and again. This is what sets life apart from nonlife: the ability to re-create—reproduce—itself. There's lots of variation in nonliving nature, too: for example, rocks are all different from each other. But there's no inheritance: rocks don't evolve because they don't re-create themselves. You can even have variation *and* selection but no inheritance: for instance, picking your favorite brand of beer in the grocery store. This does not lead to the evolution of beer because you are not re-creating it. And what if you did? Suppose you are a brewer and decide to re-create the flavor of your favorite beer. Of course it ends up slightly different, but customers like this "new generation" of beer even more than its "parent." Your beer becomes hugely popular thanks to "selection" by customers—it displaces other brands, which "die out." Is this really much different from the evolution of species? As long as there's "inheritance"—even if it comes in the form of imitating beer—evolution is possible.

But the real breakthrough of Darwin's theory is this third component of the moving car: selection, the car's driver. Selection is what decides where the car is going. It is this new figure that Darwin's theory places at the steering wheel—an invisible, omnipresent natural force that permeates the world and shapes every living creature in ingenious ways. In more traditional language, this force would most certainly be called a god.

The Power of Choice

Indian snake charmers appear to hypnotize their subjects by playing a musical pipe called a pungi, although in reality the snake follows the movement of the instrument rather than undulating in a trance. One of the snakes used in these performances is the Indian cobra, also notable for a striking pattern on the back of its hood: two black-on-white circles connected by a curved line below. These circles evoke old-fashioned spectacles—hence the other name for the species, the spectacled cobra. Maybe more to the point, they also evoke eyes. And underneath, a nose, or a muzzle, or a beak, or whatever else you might expect to find between two wide-open eyes that are *looking straight at you.*

The cobra is an animal unmatched in its grace and beauty. A cobra fighting a mongoose—look it up—is a deadly dance worthy of kung fu movies. A typical cobra mostly hunts rodents and frogs, avoiding large beasts. Why does it need the spectacles on the hood? Seems odd to even ask. Clearly the pattern is there to scare away those large beasts lurking behind the cobra. Nature has very little vertical symmetry. We humans have bridges and buildings, but in the forest virtually the only thing that is vertically symmetrical is an animal whose gaze is fixed on you. And most of the time animals who fix their gaze on you are actively entertaining the idea of eating you. So, when two big eyes suddenly rise from the grass, everyone knows they should be scared, and the cobra is saved. It is a pretty obvious strategy. But there is one question about it that I like to spin around in my head.

Does the cobra know it has eyes on its back?

To me, this question is like a koan. It's not really about the answer as much as where the question itself takes you. It exposes a mental transforma-

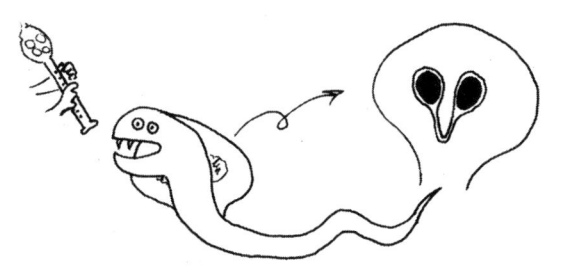

The spectacled cobra and its spectacles

tion, that breakthrough of intuition brought about by evolutionary theory. It comes right up to the edge of common sense and spits into the abyss. No, the cobra does not know about the pattern. It never looked in a mirror, let alone at its back; it leads a solitary life and is unlikely to comprehend the specific connection between itself and other cobras. But if the cobra does not know, then *surely someone must?*

Living organisms seem as if they were purpose built—full of *ideas* that must have come from *somewhere*. Plants do not think about their flowers, but surely someone must know that if you look at a sunflower in UV light, you see hidden decorations adapted for the vision of its pollinators. Someone must know that a leaf-tailed gecko looks exactly like a fallen leaf, and a stick insect like a stick. Someone must know that legs are for walking and wings are for flying, that hearts pump blood and kidneys, urine. But if not the owners of said legs and wings, then who?

According to Darwin, this knowledge arises from selection. Selection is what takes variation and molds it into shape. Selection is a reflection of the world, its current properties, requirements, and limitations in the properties and abilities of living organisms. Selection is like eternity's bouncer. It lets snakes with scary eyes on their hoods pass into the future and keeps snakes with other patterns in the past. That is why everything in nature seems so well-designed: we see only a handful of selections.

When people speak of evolution as a random process incapable of creativity of the sort that you see in the cobra's spectacles, they ignore the powerful and definitively nonrandom creative power of selection. Schoolteachers who bash into our heads that evolution has no rationality often completely ignore the fact that, in the long run, nature's creations are indistinguishable from rational designs. Sure, giraffes never conspired to extend their neck length to its present size. It wasn't a rational idea that any giraffe had. It is nature itself that had this idea, and if you think about it from nature's perspective, it is actually quite rational to give a herbivore the ability to pick leaves off tall trees. Don't call it an *idea* if you think that's too spooky or unscientific—call it an *essence*, nature's idea: a rational fruit of selection. Giraffes don't choose to be long necked—it is the leaves on tall trees that make the decision. Dolphins, seals, and tunas do not all independently decide to be gray, smooth, and oblong—it is the properties of water that bring them to these essences. Cobras have no clue about the pattern on their back—it is the brains of their enemies that slowly distill the defensive essence of two big fake eyes on their hoods. Selection is the creator of essences.

I prefer to use this more abstract term, *selection*, to Darwin's original *natural selection*, which sounds a bit outdated to my ear. *Origin of Species* is essentially built around a comparison: humans have been creating improved breeds and varieties of plants and animals for millennia by selecting desired traits in every generation and allowing only their owners to reproduce—that's called selective breeding—and, in the same way, *nature* carries out its own, *natural* selection. This was a novel idea at the time, and the comparison was needed for the popular imagination to fully grasp it. But oddly enough, *natural* here is an anthropocentric qualifier, putting the most fundamental law of life in subordination to the human practice of creating large pumpkins. To me, it seems more logical to say that the selection of pumpkins by humans is just one case of a very general principle of selection only superficially distinct from the selection of flowers by bees or the selection of fish by water.

Selection is also a simpler, more detached term, which I think highlights that Darwinism today extends much further than the evolution of species. In our modern understanding of the idea, it no longer has to apply only to living organisms competing for resources. Any time something is selected from something else based on its properties and those properties are reproduced with variation (as in the example of a brewer re-creating a favorite beer), it will evolve, and that's the point.

In a curious online experiment called DarwinTunes, users listened to audio clips generated by random variations of code and scored them 1–5. Highly rated code variants were "selectively bred," their parts copied and mixed, the resulting audio clips offered to new users, and so on. So there was variation (random code variants randomly mixed), inheritance (new code variants made of previous ones), and selection, except in this case it was based not on competition for resources, but on competition for user preferences. The result: noise evolving into music. In five hundred generations, the randomly generated wall of sound acquired rhythm, in a couple thousand, a robotic pan flute broke through, and eventually the "tunes" stabilized at about the level of a polyphonic ringtone from the early 2000s.[3]

The best, broadest definition of selection I've heard came from a friend well outside of academia: *good ideas survive; bad ideas do not.*

Who decided that it must be so? Where, in other words, does selection come from? Here's a thought experiment. Let's pretend selection doesn't exist. What would that look like? Maybe the world is inhabited by a riotous variety of fungi that constantly mutate, have zero competition with one another, and happily reproduce because they don't need much except space and some dead matter. Nothing selected over anything else! What will happen? Either the dead matter or the space will eventually run out. Competition will begin, which means something will live and something will not. Anything material has limits, and limits mean selection.

Selection is built into the logic of the universe. It cannot *not* happen. It exists by definition. Good ideas survive; bad ideas do not. What is a good idea? It is an idea that survives. What is a bad idea? It is an idea that does not. By default, everything in the world eventually disappears. But sometimes, among these things that disappear, nature allows some things to continue existing. This is selection.

If you break it down to the fundamentals, Darwin's theory only deviates from intelligent design on one account. With intelligent design, an invisible power first selects an idea and then realizes it. With Darwinism, an idea is first realized and then an invisible power selects it.

But if all that Darwin did was to replace one invisible power with another, are evolution and intelligent design pretty much the same thing—or at least equally viable theories? Not quite. Theories are explanations. To see which theory is better, we have to first understand what exactly these theories are trying to explain.

"Nothing in Biology Makes Sense Except in the Light of Evolution"

When I was in college, this title of Theodosius Dobzhansky's classic essay seemed to me to be an overexcited exaggeration, like "Babushka's borsch is the best in the universe" or "There's nothing scarier than Black Friday." I fully grasped what Dobzhansky was saying only when I began assigning it to my own students. It's not that the intervening years endowed me with some special wisdom. The reason is that I studied in Russian, but I teach in English.

It is a curious case of linguistic mismatch. The famous geneticist and evolutionist Dobzhansky was born in Ukraine but spent most of his life and career in the United States. His 1973 essay was originally written in English, so the Russian version I read in school was a translation. The titular phrase *nothing makes sense* became *nichto ne imeet smysla*. In Russian, it can be interpreted in two ways: "nothing has meaning" (as in, it is futile, impractical, or otherwise not worth your time), and "nothing has logical coherence" (as in, it doesn't add up). What I heard was the first interpretation, and it sounded to me as if Dobzhansky directed me, an aspiring undergraduate, to become an evolutionary biologist. If I wanted to, say, develop pharmaceuticals, he called that "meaningless" and excluded me from the cool club of scientists.

But in English, "to make sense" is much closer to the second interpretation, "to add up"—to have internal logical coherence. This is what the essay is about: that in biology nothing adds up unless you imagine that it originated in the process of evolution. It just doesn't make sense any other way.

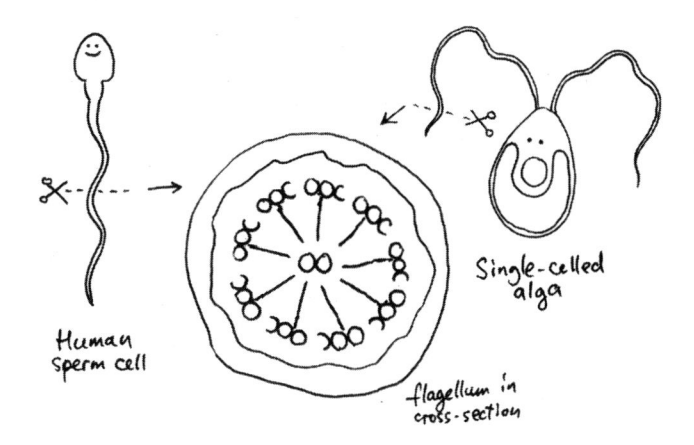

Living nature is indeed a very odd thing if you forget all biological theories and pause to consider it for a minute. Remember the Galapagos finches? It is of course possible to imagine that God created them in "centers," carefully crafting each species to be slightly different from the rest, adjusting beaks to meet equally subtle environmental differences, and combining the birds with other meticulously fine-tuned species. It just sounds a bit strange. Everything else is strange for similar reasons. Why does the world need so many species? There are literally millions of them, each one unique in its own way. Most of this planetary biodiversity remained unseen by anyone before the invention of the microscope. What is the point of that? *And why, despite their innumerable differences, are all these species nevertheless so similar?* If they were spawned by the Creator's limitless imagination, why are they all made of cells like some mad Lego project? Why do all vertebrate animals have exactly one head, two eyes, and four limbs? Why does the cross section of the human sperm's flagellum look exactly like that of a single-celled alga?

This is what the theory of evolution explains better than intelligent design. Dobzhansky in his essay puts it in two words: diversity and unity.[4] The existence of an unprecedented, unfathomable quantity of forms *in combination* with a unified orderly system of groups and similarities. Diversity and unity is what doesn't make sense in living nature, unless you allow that this living nature exists in a state of constant evolution.

To understand life on Earth, you have to bend your intuition one way or another. The easiest way to do it is to imagine God, the intelligent Creator. This explains well why everything alive is so cleverly designed. But intelligent design is not good at explaining diversity and unity. (For instance: why does the Creator need so many beetles, and given such passion, why do they all have to be six legged?) These properties of nature are explained much more intuitively by evolution, descent with modification.

Why are there so many species in the world? Because everything alive is in constant motion in all directions, and today's species are but a momentary cross section of this motion.

Why are they so similar to each other? Because they descend from common ancestors, and the closer the common ancestor, the greater the similarity.

Why are they so well-designed? Because they have been changing for a very long time and have solved so many problems that it appears as if those problems never existed.

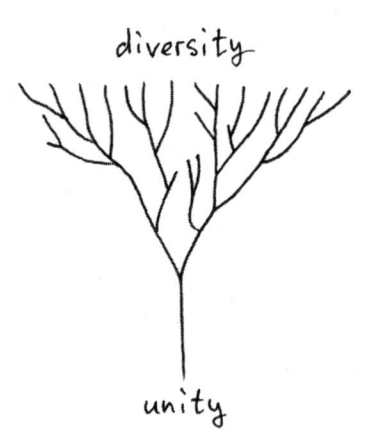

After Darwin

If we look at Darwin's theory at the scale of the cosmos, it signifies life on Earth becoming self-aware.

In the mid-nineteenth century, when *Origin of Species* first came out, the idea of biological evolution—meaning changes to a species over time—was widely unfamiliar and religiously risqué even among educated elites. Darwin was not the first person to come up with the notion that species can change. But he did provide an explanation for this idea that people not only believed but saw right before their eyes.

Today, the idea of a species undergoing gradual change comes as no surprise to any scientist, and it has even entered the popular imagination. You can see evolution. You can see it in the emergence of virus variants. You can see it in the breeding of animals. You can see it in the geological record: the transformation of fins into hands, for instance, has been reconstructed in great detail. You can see evolution most clearly in laboratory settings: microorganisms and even human cells evolve right under the microscope, and a method called *directed evolution* is a routine part of biotech.

From our current viewpoint, Darwin's theory is of interest for a different reason. It doesn't just explain how one species descends from another. It explains how *all* species descend from *one*. It doesn't matter which species—human, bird, snail, or mushroom—in Darwin's interpretation, they all turn out to be relatives, parallel plotlines of the same story that begins at one specific point of an unimaginably distant past. In the never-ending race against extinction that has been going on ever since, all surviving species are the current leaders.

Before Darwin, humans had been the highest form of life. Thereafter, they became just one of life's many branches. Previously, the world had been static, and the diversity of living forms simply existed in the way the Creator had planned. Now, the world has become dynamic: each creature, event, and happenstance connected through chains of causality that traverse thousands of generations, converging branch by branch into the trunk of the tree of life, linked to a single universal source.

This was a revolution in the relationship between man and nature and, to be sure, in the relationship between man and God.

But to see the Darwinist vision of the world as lacking rationality or creativity of the kind that only a God could offer is to utterly misunderstand evolution. Darwin's nature as divine and delightful as nature has always been—just more understandable.

To quote the British biologist Richard Dawkins: "Intelligent life on a planet comes of age when it first works out the reason for its own existence. If superior creatures from space ever visit earth, the first question they will ask, in order to assess the level of our civilization, is: 'Have they discovered evolution yet?'"[5]

Chapter 3.
The Birth of Complexity

> And this elaborate machine
> I made myself, from barley.
>
> DANIIL KHARMS

It is an interesting twist of history that the tree of life, a Biblical image, was appropriated by evolutionary biologists to represent relationships between species. Although Darwin meant no insult, religious conservatives derided the evolutionary use of the term ever since. But early in the Darwinist era, evolutionary "trees of life" bore a distinct imprint of their Christian and even ancient Greek roots. A good example is a famous 1879 "tree of life" by the German scientist and artist Ernst Haeckel, a key popularizer of Darwin's theory.* Haeckel's tree is tiered: at the bottom of the trunk he positions "protozoa," referring broadly to many types of microorganisms; above them, "invertebrate animals" with various branches of worms, insects, and mollusks sticking out left and right; above, "vertebrates," including fishes, birds, and reptiles; and, finally, "mammals" on top, crowned with "apes" and, at the very tip of the tree pointing upward, "man." Early evolutionists accepted that all species were related, but they didn't immediately let go of the notion that humans were, somehow, number one.

---

*He also popularized scientific racism, based on presumed patterns of human migration from the Garden of Eden. It was an era full of intellectual contradictions.

It is, in fact, a trope in human culture, from Aristotle to *2001: A Space Odyssey*: humans are an especially perfect species chosen by divine or cosmic forces to rule the world. Ancient Greeks, medieval scholastics, and early Darwinists had different conceptions of what life on Earth was about, but one idea they shared was a hierarchy of being: a natural progression of perfection, from rocks to plants to beasts to humans, and then, optionally, to tiers of gods or angels. Haeckel's evolutionary tree bears the same message: some living creatures are more perfect than others, and we humans are the most perfect of all—second only to gods.

Over the following century, as the implications of the evolutionary theory slowly dawned on humanity, this human exceptionalism became harder to justify. If all species come from the same common ancestor, then everything alive today has been evolving for the same amount of time and with the same result: we are still here. So who's to say humans are the most perfect? Today's evolutionary theory is blunt on that account: all living species are coequal winners in the brutal game of life, and no one is "more perfect" or "more evolved" than anyone else. There is no standard by which we can claim to have evolved better than, say, sulfur-reducing bacteria or baobabs.

But this leveled playing field challenges our thousand-year-old intuitions. Aren't humans the most advanced, most powerful, most sophisticated creatures on the planet, the only ones who can build skyscrapers, control nuclear fusion, and play in symphony orchestras? Maybe we are no longer the crown of creation, but surely there's *something* special about us?

I think both notions can be true. Yes, we are just one evolutionary branch among many, and in no way are we more perfect or better evolved than others. But that doesn't mean we are not special. Each branch of the tree of life has its *thing*, its essence that makes it unique, sets it apart from others, carves out a niche of existence among the infinite ways to be alive. Jellyfish have stinging cells with no equivalent in the animal kingdom. This doesn't make them "more perfect" than anyone else, but it sure makes them special. We also have a thing that makes us special, if not necessarily more perfect. *Perfection* is too loose a term, and it carries an unnecessary undertone of achievement. A much better, more neutral, more specific term that describes the essence of the human species is *complexity*. As we will see, complexity is not something that everyone automatically wants. But it is *our thing*. We humans have a love affair with complexity. In fact, it started long before the birth of the first Homo sapiens, before even the advent of the animal kingdom or any life on land, billions of years ago, in the depths of the primordial ocean. To understand the essence of our species, we must first understand something much more ancient: the essence of our domain of eukaryotes.

The Strange Domain

When genetic sequencing first developed in the 1970s, scientists were suddenly faced with a titillating possibility: you could now *calculate* the tree of life. If you compared the same genes of several species and measured their similarities and differences, you could work out the relationships among species—this, in a nutshell, is how all evolutionary biology works to this day. The approach was pioneered by the American scientist Carl Woese, who identified three major branches, or domains, of life.[1] The three domains were eukaryotes (our own domain, which includes all animals, plants, and fungi), bacteria, and archaea. This last one was somewhat a surprise: archaea were previously considered to be a subtype of bacteria, known to favor extreme habitats such as thermal acidic pools (although some also inhabit more conventional places, like our mouth). As it turned out, archaea and bacteria have been evolving separately from almost the beginning of life on Earth and represent two distinct evolutionary stories. Even more surprising, eukaryotes turned out to be more closely related to archaea than to bacteria—more on that later.

As with any modern evolutionary tree, all species on Woese's tree are shown at the same level, a big departure from Haeckel's tiered branches. No domain and no species is more evolved than any other.

But just as it is hard to let go of the notion that humans are somehow the number one species, it is also hard to let go of the notion that despite all the kumbaya, eukaryotes must somehow be the number one domain. Even the name implies that much—"eu" means "true" (the other part of the word, "karyon," is Greek for "nucleus," a feature of a eukaryotic cell.)

Just look at these three domains! All archaea and bacteria* look more or less the same. They are all boring, bubblelike, single-celled microorganisms. There's

---

*Collectively they are referred to as "prokaryotes"—basically, "not quite the real deal."

Carl Woese's Tree of Life

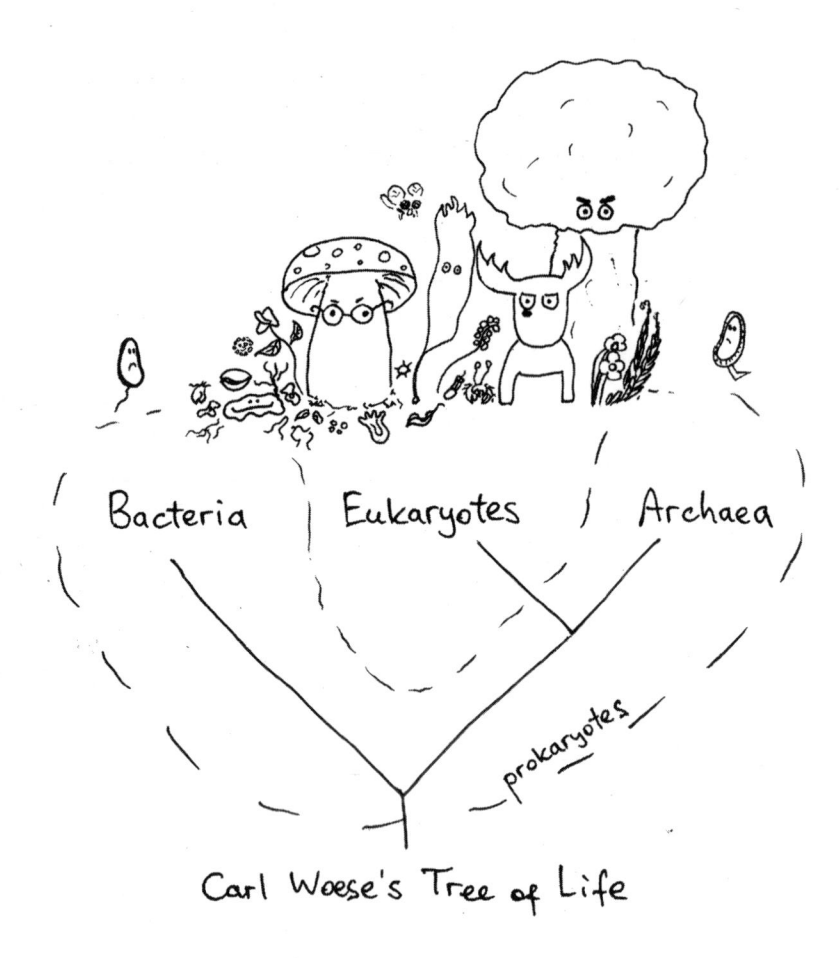

no multicellularity, no mating rituals, no symbolic communication systems, nothing that we eukaryotes consider the measures of greatness in living nature. Our own domain, by contrast, teems with that kind of stuff: legs, twigs, beaks, teeth, nectars, venoms, languages, dances, colors; the bizarre geometric shapes of microscopic diatom algae; the drumming display of a jumping spider; the putrid smell of the rafflesia flower; the cordyceps mushroom controlling an ant's brain—the eukaryotic domain is where nature's creativity seems to shine in ways both delightful and macabre but on a level seemingly inaccessible to other domains.

It's not like bacteria and archaea are pushed to the sidelines of life on Earth. By sheer biomass, for example, there's thirty-five times more bacteria in the world than there are animals, and even the comparatively exotic archaea outweigh the animal kingdom more than three times.* There's a lot of them, and there's also a lot of different kinds of them—actually, by their overall genetic diversity both bacteria and archaea are far ahead of eukaryotes. The bacteria that cause food poisoning and the bacteria that produce yogurt are a lot more different from each other than humans are from bananas.

So if there's so many of them and none of them does anything remotely interesting, what are we eukaryotes doing right that none of these multitudes of bacteria and archaea can figure out?

You can tell that the puzzle with our domain is basically the same as with our species. It seems obvious that we are somehow better than everyone else. But evolutionary biology tells us there is no such thing as "better." These two puzzles—eukaryotes versus the rest and humans versus the rest—are actually the same, and in both cases, the solution is complexity. We are not better. We are just ridiculously complicated.

As we are about to see, there's a distinct thread that connects the birth of the eukaryotic domain with the birth of the first humans. It is our first eukaryotic ancestors that began the love affair with complexity that would eventually, eons later, produce the human species. It would take another two billion years before the first Homo sapiens would walk the Earth, but as an idea, an essence, we humans begin here, approximately halfway between the origin of life and the present day, when the first single-celled eukaryotes burst into the ancient ocean previously inhabited by bacteria and archaea. Let us set the stage for their arrival by reviewing the state of living nature at this point.

The Cyan Revolution

Besides life itself, the greatest thing ever invented may have been photosynthesis. By planetary standards, it appears to have followed the advent of life on Earth very quickly, almost immediately.[2] The credit for this invention goes to the domain of bacteria. It is hard to come up with anything more consequential to the history of the planet and for multiple reasons.

---

*Eukaryotes do have a greater biomass than bacteria and archaea overall, but this is largely due to trees—just think of all the wood in all the jungles and taigas weighed together.

The first reason is food. This is what's *synthesized* (assembled) in photosynthesis. To fully grasp the importance of what this means, consider a sandwich. You can consider a BLT if you prefer, but I will be considering the revered Soviet classic, *buterbrod s ikroi*, an open-faced sandwich with salmon caviar. It consists of puffy white bread, a generous smear of butter, and salted fish eggs. What is the bread made of? It is made of wheat (so is vodka, which is as mandatory a pairing as wine is with cheese). Wheat is a plant. It was made by photosynthesis. What is the butter made of? It is made of milk, which is made by a cow. The cow is made of the grass that it ate, and the grass was also made by photosynthesis. What is the salmon roe made of? It is made by a mother fish, and she is made of smaller fish and crustaceans that were her prey, and those, of the algae that they ate, and so the caviar is also ultimately made by photosynthesis. The only part of the sandwich that is not is salt.

Photosynthesis doesn't just make food for plants—it makes food for everyone, ultimately feeding almost every living creature on the planet. During photosynthesis, energy from the sun is captured and used to cobble small, lifeless molecules of carbon dioxide into large, organic molecules of sugar. The energy from the sun is injected into the bonds between atoms that form the scaffold of the sugar molecule, holding them together. It is not critically important that it's sugar, specifically: once you make that, you can turn it into other organic molecules, like fats, or proteins, or DNA. But any time you transform molecules from one form into another, you lose a little bit of energy stored in the bonds between their atoms. So the energy must initially come from some external source. Almost universally, this original source of energy for life on Earth is the sun. Exceptions exist—there are, for example, bacteria that feed off chemicals coming from hydrothermal vents—but on a planetary scale, they pale by comparison. It would not be an exaggeration to say that life as we know it is powered by photosynthesis.

If this is not enough of a game changer, there is a second reason for why photosynthesis is the most significant thing ever invented: oxygen.

Oxygen was not actually part of the original, most ancient forms of photosynthesis.[3] For technical reasons, the process of harvesting energy from the sun requires, besides carbon dioxide and light, a source of electrons—negatively charged particles that hover like clouds around atoms and link them into molecules. These electrons must be extracted from some additional donor molecule. This donor molecule is split, the electron is removed, and a byproduct remains behind. The byproduct depends on the donor substance used to harvest the electron. Originally the substance may have been iron,[4] hydrogen,[5] or hydrogen sulfide,[6] all of which are excellent sources of electrons and leave behind benign byproducts—for example, hydrogen sulfide leaves behind sulfur. But whether you take electrons from iron or hydrogen sulfide, it means that you depend on these substances as resources. This dependence made the oldest forms of photosynthesis not quite as magical as photosynthesis is today. Sure, you could make food out of thin air and light—but you still had to find some iron- or sulfide-rich geyser to do that and sit next to it forever. Various types of bacteria apparently did that for hundreds of millions of years.

Everything changed with a new source of electrons—water. The switch to water as a donor molecule was pioneered by a group of upstart organisms known today as cyanobacteria. It required an astronomically complex molecular machine with a rather uninspiring name, Photosystem II. What this gargantuan protein complex was able to accomplish was something previous versions of photosynthesis didn't—it split the water molecule and extracted from it the requisite electron. This changed the rules of the game. Now, instead of relying on scarce resources like hydrogen sulfide or iron, cyanobacteria could run their photosynthesis on one of the most abundant chemicals on the planet. The switch to water untethered cyanobacteria from almost any worldly dependence. Food was now literally made out of air, water, and light—truly closer to magic than to chemistry. This revolution in photosynthesis made cyanobacteria some of the most prolific organisms on the planet. But it had an unexpected side effect. As water became the donor molecule for electron extraction, the byproduct of the process changed. When hydrogen sulfide is broken down, it releases sulfur, which is relatively inert. When water is broken down, it releases into the environment something a lot more potent: oxygen.

You might recall from chapter 1 that oxygen is the Shiva the Destroyer of the atomic-molecular world, an aggressive element that violently attacks large molecules and breaks them down into the smallest bits, releasing massive amounts of energy—this is how combustion works. So, by its nature, oxygen is toxic.*

Once water-based photosynthesis was invented, oxygen started to build up, first in the ocean then, eventually, in the atmosphere. Soon the majority of

---

*To be precise, what's released as a byproduct of photosynthesis and consumed during combustion is *molecular* oxygen, O_2, two atoms of oxygen that join hands on their way to attacking other molecules.

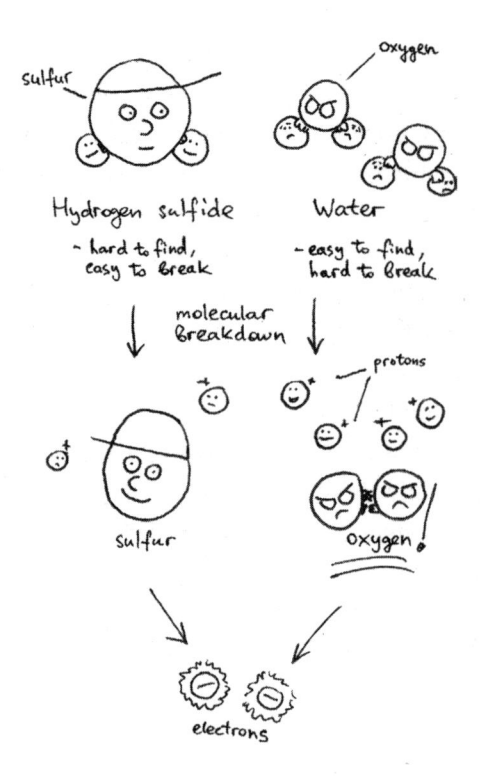

[Figure: Hand-drawn diagram showing molecular breakdown. Hydrogen sulfide (sulfur) — "hard to find, easy to break" — and Water (oxygen) — "easy to find, hard to break" — undergo molecular breakdown into sulfur, oxygen, protons, and electrons.]

bacteria and archaea alive at that point could no longer tolerate this much oxygen. Many of them—most, probably—died and were forever lost to history.

This was the world as it stood at the time when the third—eukaryotic—domain was about to come to life: a world whose very existence was threatened by a dangerous chemical shift on a planetary scale. Some scientists call this moment the "Great Oxygenation Event," the "Oxygen Catastrophe," or even the "Oxygen Holocaust"—the latter term literally translates as "all burning," which, given oxygen's violent chemical properties, is a literal description: life on Earth slowly burned in the toxic waste of photosynthesis.[7]

There was, however, a glimmer of hope: respiration. Just as pathogenic bacteria today evolve antibiotic-resistant strains when confronted with penicillin, various strains of bacteria and archaea billions of years ago evolved resistance to oxygen that was destroying the rest of their kin. This is what respiration is—a way to withstand oxygen's destructive power. To do that, you need to give oxygen something to destroy—some spare organic molecule, a piece of food.

The good news is that, once you find that piece of food to throw into oxygen's bottomless gullet, the appeasement comes with an added bonus: you get more energy in return. Because oxygen is so good at ripping molecules apart, once you learn how to channel its power, you can use it to squeeze out every last bit of energy from whatever molecule you hand it. Basically, respiration is controlled combustion. The "controlled" part keeps you from burning up yourself, and the "combustion" part extracts maximum energy from your food. You can break food down without oxygen—it happens, for example, in our muscles during intense exercise—but doing it with oxygen yields up to *nineteen times* more energy. So it's not just a small bump—spicing your food with oxygen increases your energy supply by more than an order of magnitude.

But what kind of food is available? Cyanobacteria make their own, thank you very much. But for other, non-photosynthesizing organisms, the options are limited. There are not yet any plants to decompose, no animals to invade, no pickles to ferment, no soup to spoil. Their best bet is probably scavenging at the bottom of the ocean for scraps of dead cyanobacteria, and so breathing bacteria graze the seafloor for this algal scum. Not the most appetizing lifestyle.

The stage is set for the arrival of eukaryotes. It is time to finally meet the direct ancestors of the eukaryotic domain. Our quest leads us to Asgard, home to the Nordic gods.

Loki's Castle

In 2015, computer scientists from Uppsala University discovered an organism that appeared to fall right at the border between the domains of archaea and eukaryotes. No one had ever seen it—the discovery was made by analyzing DNA fragments extracted from a soil sample. The sample had been collected five years earlier from sediment surrounding a hydrothermal vent located between Greenland and Norway. The vent is called Loki's Castle, in reference to the region's Nordic mythology.[8] Scientists decided to name their newly discovered eukaryote-like archaea *Lokiarchaeum*, "Loki archaea," based on the source of the sample.

Their discovery was published to great fanfare in the prestigious journal *Nature*. Then, within just a few months, another group of scientists from the United States discovered another similar group of archaea from a different source, which they named, for the hell of it, Thorarchaeota, in honor of another Nordic god and superhero.[9] There was no going back. Next came Odinarchaeota and Heimdallarchaeota, and, as you can tell, no one is about to stop the

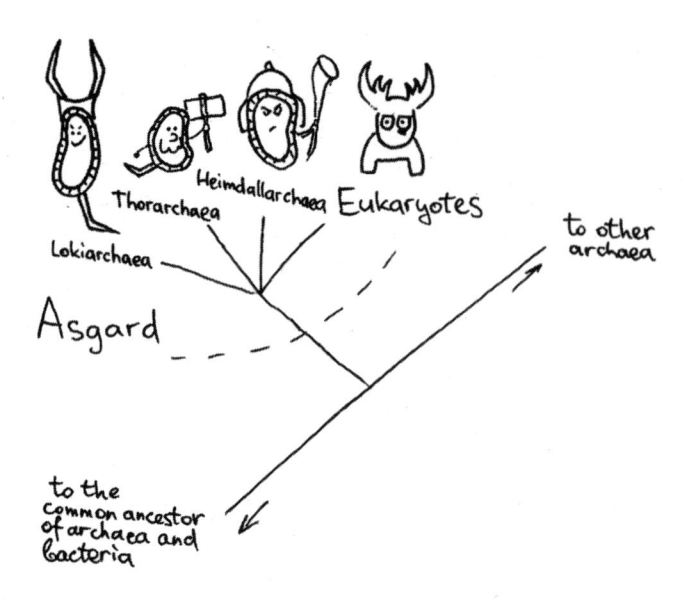

[Figure: phylogenetic tree labeled with Lokiarchaea, Thorarchaea, Heimdallarchaea, Eukaryotes, grouped as Asgard, branching to other archaea and to the common ancestor of archaea and bacteria.]

party. This whole bunch is now known as Asgard, the mythical world where all these gods hang out.[10] These creatures are the closest thing among prokaryotes that there is to eukaryotes. In fact, the most recent evidence indicates that eukaryotes evolved *within* Asgard, and so, technically, we could consider ourselves Asgardians too.[11]

Finally, in 2020, the world glimpsed its first look at Asgard archaea—until this point we knew of them only as strings of DNA code. It took researchers from Japan twelve years to cultivate them—a truly Japanese level of commitment.[12] This time, the elusive organisms were sourced from a methane seep, a patch of the ocean floor rich in methane and free of oxygen—one of the few corners of the planet where nonbreathing organisms can still find respite from oxygen toxicity. The scientists called the species they managed to cultivate *Prometheoarcheum*, in reference to Prometheus—a transitional form between gods and humans, I suppose, or between Nordic and Greek mythology, anyway. As a result, Prometheus is now officially one of the subtypes of Loki archaea.

Mythological confusion aside, what exactly made Asgard archaea so special? What their genes revealed was a primitive version of a *moving membrane*. And although neither Prometheus nor other Asgard archaea surviving to this day ever quite realized the potential of their invention, one branch among them did. We, eukaryotes, figured out that a moving membrane can be used to *eat other living organisms whole.*

48 One Hand Clapping

Power Grab

Bacteria are capable creatures, but one thing they strangely cannot do is eat one another. They can't bend their membrane in such a way so as to swallow another cell.

A membrane is a key element of any living cell—a thin, perfectly sealed capsule made of specialized fats. A membrane does not simply open—for most cells, a hole in a membrane means instant death. So you can't just push a large piece of food—such as, for example, another living cell—through the membrane. Only small molecules, such as sugars or amino acids, can traverse the membrane, and this is what bacteria and archaea eat if they don't photosynthesize. They can deter, poison, trick, and outcompete each other, but they cannot swallow each other whole.*

Our cells, on the other hand, are very good at such things. When a rogue bacterium invades the human bloodstream, it gets attacked by an immune cell called a macrophage. From the bacterial perspective, the macrophage is a giant monster that engulfs the bacterium in its membrane. A vesicle (membrane bubble) containing the trapped bacterium buds off inside the macrophage's membrane. Next, the vesicle fuses with another, weaponized vesicle called a lysosome, an organelle packed with digestive enzymes, which upon fusion with the bacterium-containing vesicle dissolves the microbe alive. The macrophage absorbs the nutrients, which pass across the membrane of the vesicle into its own cytoplasm, and it resumes sniffing out trespassers. This membrane trickery solves the problem of getting a large piece of food inside a cell: rather than pushing it through a hole, it is engulfed in an internal bubble.

All this engulfment and fusion of membranes—collectively known as *vesicular transport*—looks simple, natural, like the movement of oil droplets. Eukaryotic cells bustle with such movement, packed with semiliquid, undulating membrane compartments of all shapes and sizes. In fact, controlling such oil-like flexibility with seamless precision requires extraordinarily complex molecular technology. The flexible cell membrane must be suspended on a moveable cellular skeleton, or cytoskeleton. There must be proteins that control membrane curvature, membrane budding, and membrane fusion, proteins that

---

*As always, there are exceptions, and "swallowing whole" has recently been discovered in one bacterial species.[13]

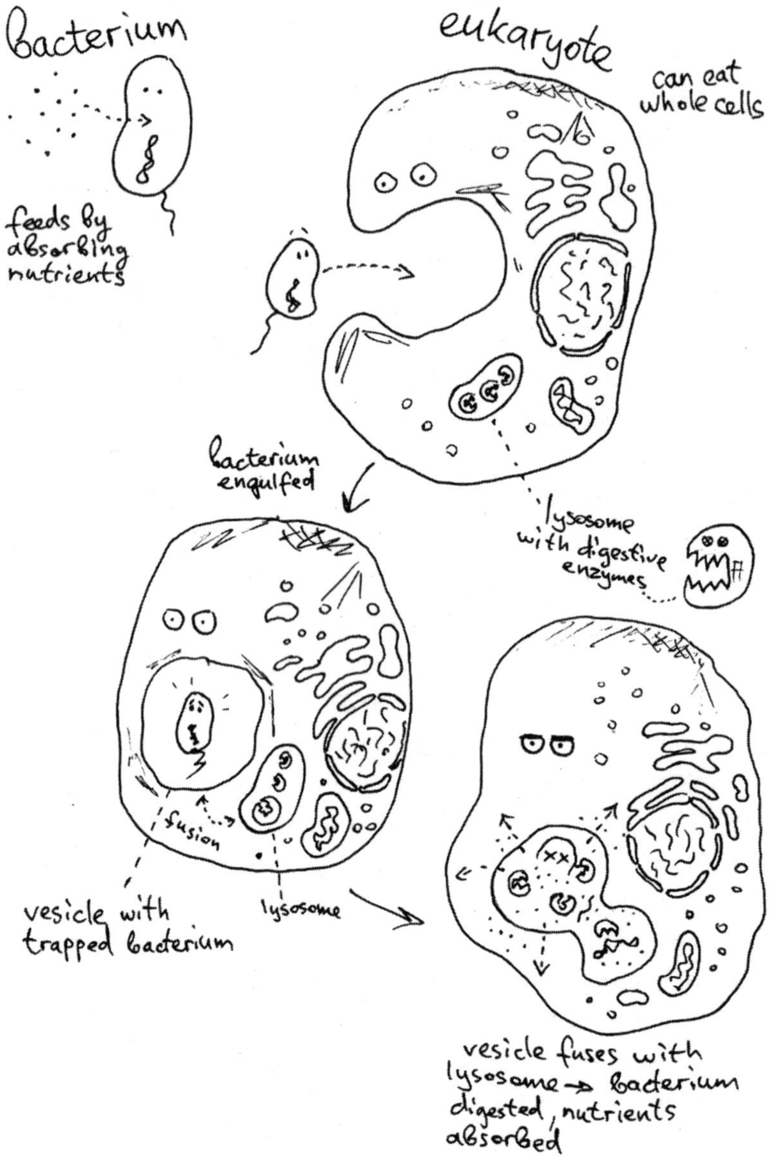
bacterium
feeds by absorbing nutrients
eukaryote
can eat whole cells
bacterium engulfed
lysosome with digestive enzymes
fusion
vesicle with trapped bacterium
lysosome
vesicle fuses with lysosome → bacterium digested, nutrients absorbed

drag vesicles and organelles around the cell, proteins that lay out the rails for those proteins to move, and proteins that take those rails apart.

It is this kind of machinery, in primitive form, that was found in Loki archaea, as if they were one step away from becoming eukaryotes but never quite got there. The shape of Prometheus, the live Asgardian that Japanese scientists were able to cultivate, also suggests a moving membrane: its cells resemble an amoeba or octopus—a loose ball with flexible arms protruding in all directions. These arms, in fact, were key to Prometheus's survival: it refused to grow unless supplemented with another organism, a metabolic comrade, which it cuddled in its soft tentacles. Only in partnership with this additional creature, which aided it in digesting nutrients falling onto the seafloor, was Prometheus viable.

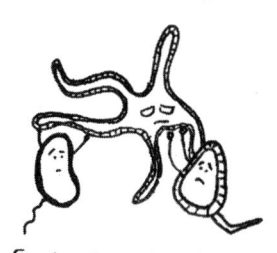

Feeble Prometheus and his metabolic comrades

Suddenly, the pieces of the puzzle of eukaryotic origins started falling into place.

One day, about two billion years ago, as the world still reeled from the effects of the Oxygen Holocaust, two creatures met each other. The encounter must have happened at the border between the ocean's water, which by now was saturated with oxygen, and some oxygen-free haven such as a methane seep—so perhaps on the seafloor. One of the creatures was a breathing bacterium. It didn't fear oxygen because it mastered respiration and was looking for nutrients to burn. The other creature was an Asgard archaeon. It couldn't breathe and so was restricted to living off seafloor detritus—and even for that, if Prometheus is any indication, this feeble creature needed help from some additional metabolic comrades. The breathing bacterium, in fact, would have been a perfect comrade—what better to help with digestion than an organism that can control fire? Using its flexible membrane, the Asgard archaeon drew the breathing bacterium into its orbit, pulled it close, wrapped its tentacles around it, tightened the grip, fusing the tentacles together, and then, suddenly, found the breathing bacterium inside of itself.

This was the moment that created our domain. A new compound organism consisting of an archaeon and a bacterium was born. This double origin meant it could do two things: engulf other cells, which was handled by the moving membrane of the archaeon, and burn them up into molecular shreds using oxygen, which was done using the bacterium's respiratory abilities. As a result, this dual cell gained access to quantities of energy that individual bacteria and archaea could only dream—*it could take energy away from others*. Out of a friendly but feeble bottom feeder and a hardened survivor of the Oxygen Holocaust was born a new, magnificent, and terrifying creature capable of destroying other cells whole and thriving on their ashes: the Death Star of cellular life, a eukaryote.

The bacteria that gave eukaryotes the ability to breathe are still with us, inside each of our cells. Today we call them mitochondria. They still have their own genes and still divide semi-independently of the cells they inhabit, as if for them, our bodies are just a highly specialized environment as good as any. Think about it: there are ancient organisms inside all of your cells, and they are doing your breathing for you. There's another Star Wars reference here: when George Lucas was creating the Jedi, mitochondria served as an inspiration for midichlorians, fictional microscopic creatures that give Jedi their powers.

Although human ancestors were not part of this particular journey, there actually has been a sequel to the birth of eukaryotes: another round of organisms fusing and merging their abilities. What I am talking about is the origin of the kingdom of plants. Just as the first eukaryote swallowed a breathing bacterium and turned it into a mitochondrion, the ancestor of plants, already a eukaryote with mitochondria, additionally swallowed a cyanobacterium—that oxygen-spewing photosynthetic organism—and turned it into part of itself, now called a chloroplast. These formerly free-living chloroplasts are what makes plants green. They perform photosynthesis for plant cells to this day.

So, technically, all breathing and all photosynthesizing in the world is still done by different kinds of bacteria, even if they live inside the cells of plants and animals.

Life in First Person

The birth of eukaryotes is so significant not just because it created the largest, most complex, most energetically expensive organisms that existed to that point. What is even more significant is that it broke the mold of the bacterial way of life. Bacteria operate not as individuals, but as populations, or strains. Eukaryotes, for the first time, begin operating not only as strains, but also as individuals.

In class, I show students a cartoon in which a grumpy kid microbe sits at a dinner table and the mom microbe beside him says, "But Timmy, you have to eat your antibiotics, or you'll never become a big and strong bacteria."[14] The cartoon is funny, but it's wrong. Timmy is not *bacteria*. Timmy is *a bacterium*. *Bacteria* is plural. Timmy himself will not actually become bigger or stronger by eating antibiotics. He will most likely die. But as long as Timmy has billions of siblings, there's bound to be one that can withstand the antibiotic. That lucky sibling will be the one to survive and multiply. As a result, Timmy's entire kin becomes stronger—resistant to the antibiotic through the brute force of selection. It's basic Darwinism.

Although bacteria were first discovered in the seventeenth century, no one knew what to make of them until the advent of the germ theory of infectious disease in the late 1800s. Then, suddenly, scientists announced that not only was the world overrun with hordes of tiny, unseen creatures, but apparently it was them who had been making people sick all along. Understandably, the public's gut response was panic: at the turn of the twentieth century bacteria were blamed for every possible human ailment including aging itself, and newspapers were full of snake oil remedies and disinfectant solutions claiming to facilitate their total eradication. When penicillin was first discovered, it was seen as a miracle cure, a triumph of the human mind over the microbial forces of disease—finally, we found a way to kill them all, for good. And indeed, no major epidemic since the discovery of antibiotics has been caused by bacteria (such as plague), while life expectancy in developed countries continued to grow seemingly without limits.[15]

But we misunderstood—and underestimated—bacteria. Bacteria are not like tiny humans or even tiny bugs. They are much more formidable—an amorphous, flowing mass of genes packaged into cells, capable of solving any problem they face. They are more like the liquid cyborg T-1000 from *Terminator 2*, the one who can turn into quicksilver and move through narrow spaces. Shortly after penicillin went into mass production—it started during World War II—doctors began reporting patients who were insensitive to the drug: they were apparently infected with bacteria that had evolved resistance to the antibiotic. Today, penicillin is rarely used for any serious infection because penicillin resistance among bacteria has become commonplace. Worldwide we are, in fact, running dangerously low on antibiotics that still work reliably and might soon find ourselves in a post-antibiotic era when a simple scratch could be life-threatening.[16] Bacteria are catching up.

Antibiotic resistance is just one example, but this is how bacteria operate in general. They evolve their way around any problem, no matter what you throw at them. Their organisms are so simple and they reproduce so fast that they can bounce back from catastrophic devastation in a matter of hours. It is extremely difficult to kill them.

It is hard for us to understand bacteria because we humans associate ourselves with our organisms. When we say, "I," we mean a material object with a voice, a hair color, and some genes contained inside. Bacteria don't invest in their organisms nearly as much. Bacteria think in groups, strains, branches of the evolutionary tree. If they could say "I," they would refer not to the organisms containing genes but to the genes traveling through time in replaceable organisms.

If you look at bacteria (and archaea) this way, it starts to make sense why they all look the same. From a bacterial perspective, organisms must remain as cheap and dispensable as possible. Bacteria are not interested in complexity. They must think of all our eukaryotic peacocking as rather daft. They keep their strains alive through constant, rapid, flexible evolution of their simple organisms.

On the other hand, eukaryotic organisms, from the very beginning, were precious. Thanks to their ability to swallow and consume other cells, eukaryotes were more complex than bacteria or archaea could ever be. But their mega-

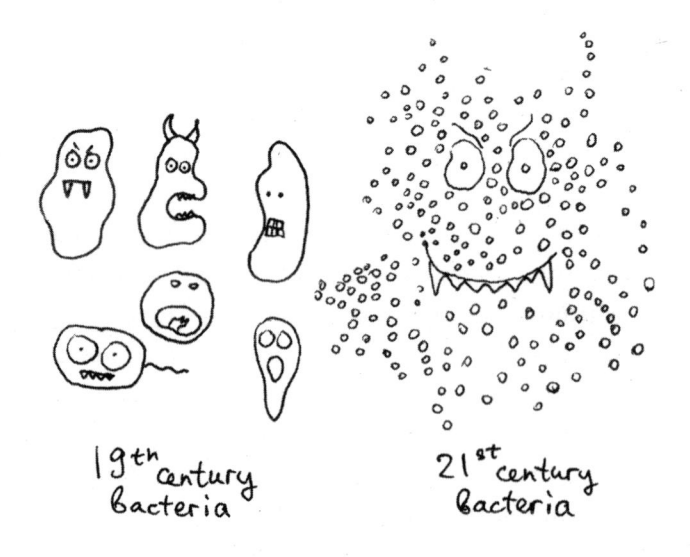

19th century bacteria

21st century bacteria

54 One Hand Clapping

cells depended on an unprecedented supply of energy to sustain themselves, to swallow and not be swallowed. And once eukaryotes started investing in these complex organisms, the entire course of their evolution went into territory uncharted by any prokaryote.

How do you ensure that your precious organism survives under the constant pressure to swallow and not be swallowed? There are many ways—defense, offense, tolerance, specialization. Each branch of eukaryotes picks its own way. Some eukaryotes grow big teeth; others grow thick shells. Some learn to survive without food for a long time; others learn to find food in places where no one else can.

But one strategy that always seems to be an option is to double down on complexity. To ensure that your investment—the mega-organism—is protected, you could choose to invest even more and make the organism even more mega. Even bigger. Even stronger. Even smarter. From the bacterial standpoint, it's like gambling away your fortune on roulette. And indeed, all their expensive innovations make eukaryotes quite fragile as strains—dinosaurs die out a lot easier than, say, staphylococci. But eukaryotes always find among themselves someone who simply refuses to step away from the roulette table, someone who *does* double down on the complexity bet and keeps going further, who adds yet another element to its organism and by doing so finds a new way to interact with the environment, to exploit it for more energy—despite the risks of extinction. As time goes on, more and more complexity evolves to tap into more and more energy, even as stakes continue to grow.

Do you see where I am going with this? The birth of eukaryotes set in motion a chain of events that finally—almost predictably—led to the birth of the human species. Eventually, something like us humans—complex enough to extract energy from fossil fuels and even atoms but still hungry for more, even as we teeter on the edge of self-annihilation—was bound to happen.

I think *complexity* is a much better term for what the medieval scholars called "perfection"—whatever it is that makes humans second only to gods. It is not about physical power (alligators are stronger than humans) nor about resilience (worms can survive in acid), nor even about the overall impact on the world (cyanobacteria were unmeasurably more consequential). It's about how complicated our interaction with the world is—how many different things our bodies and brains can do within it. If we use complexity as a measure of perfection, then eukaryotes are certainly more perfect than bacteria, and humans are indeed the most perfect creatures of all. But there is no objective reason why perfection should be defined in this way—being complex is not the only way to be. Complexity is just one essence among many. But it is an essence at which we uniquely excel, as a domain and as a species.

Complexity in exchange for copious energy use—this is the story of eukaryotes as a whole and of humans more than anyone. Nowhere in nature does this story reach quite such a climax as it does inside the human brain. Our enormous, warm-blooded bodies are already cranked up in terms of their fuel consumption to almost the biochemical maximum, but the brain uses ten times more energy per unit of weight than the body on average. This level of energy use is only possible thanks to oxygen that constantly burns through totally unreasonable quantities of fuel. Most of the energy thus released is used for maintaining the electrical charge on the membranes of a hundred billion neurons.[17] If we stop breathing, within a few minutes the neurons in our brains run critically low on energy, even though they can still break down fuel—just not as efficiently. The charge on those membranes soon begins to drop, which ultimately triggers an uncontrollable release of neurotransmitters, leading to convulsions and death.

Humans have no greater addiction than to this toxic gas of complexity that at one point almost destroyed life on Earth: a few minutes without oxygen, and our brains fry themselves. Isn't that ironic?

Chapter 4.
When All Else Fails

> It requires a great man to resist
> the common sense.
>
> FYODOR DOSTOYEVSKY *Demons*

It is tempting to assume that we humans have stopped evolving.

Evolution, it seems, is a brutal enterprise. To get to where we are now, untold numbers of our potential ancestors had to be rejected and discarded. For every single thing our bodies do for our survival—from the sense of hunger to the shivering reflex; from breathing to blinking—there had to be someone who died, leaving no offspring, because their body couldn't do that thing well. For there to be winners, there had to be losers.

But humans, it seems, are no longer as brutal as Darwinian evolution requires. The progress of humanity—both technological progress and moral progress—taught us both the ways and the reasons to support the weak, cure the ill, and even bring children to those who can't have them. By doing that, we surely have dampened the force of selection that would otherwise keep pushing us to become stronger and healthier as a species. It is logical to conclude that our evolution has stalled.

And yet that conclusion is short-sighted. There's a tongue-in-cheek principle in biology known as Orgel's Second Rule: "Evolution is cleverer than you are." As a biologist, you gradually understand that the rule is not a joke. For all intents and purposes, it is safe to assume that any logical conclusion about evolution's constraints—that it can't do this or it can't do that—reflects a failure of imagination. And so it is here: we haven't stopped evolving, we just fail to perceive how our evolution might proceed.*

So where, then, are we going? At least in principle, what could our future look like? Will we grow wings? Lose all hair? Develop distinct arms, one for fine tasks, one for hard labor? Will we learn to shut our ears like we close our eyes? (These were all features of a "man of the future" from a Russian children's book I used to have—the ear shutters especially would be a boon in noisy Brooklyn.)

We can't predict the future, but we can look into the past and search for what is possible. As a matter of fact, there has been a time in the history of our ancestors—the young domain of eukaryotes—when their evolution would have appeared to them similarly stalled as it does to us now.

It is almost unbelievable how similar all eukaryotes are at the level of cells and molecules, even if they look nothing alike from a distance. Look beyond surface appearances and into a eukaryotic cell, its molecular architecture, its busy protein machines hustling and bustling, its internal dance of vesicles and vacuoles moving in all directions. It is an entire world existing at the nanoscale. And almost every element of this world exists, in recognizable form, in humans, turtles, amoebas, trypanosomes, lichens, eucalyptuses, and all other eukaryotes alike. For example, there are two proteins called mTOR and AMPK that sense the cell's energy and nutrient levels. They act like two opposing forces: AMPK is turned on when energy is depleted, blocking cell growth, and mTOR is turned on when nutrients are abundant, promoting cell growth.[1] These two proteins do the exact same things in humans, plants, and even ciliates—bizarre microscopic eukaryotes that look like tiny fuzzy aliens. Another example: there's a protein that I study inside neurons called MAPK that is part of the process of memory formation—when a neuron is repeatedly stimulated, MAPK is activated using two separate chemical switches situated side by side, whereupon it proceeds to restructure the neuron, thereby helping store the memory.[2] Almost the same protein with the exact same two chemical switches exists in yeast, where it helps adapt the cells to a salty environment if the yeast accidentally end up in one.[3] (It also exists in every other eukaryotic

---

*Orgel's First Rule (named for the evolutionary biologist Leslie Orgel) is less known and more specific but equally insightful: "Whenever a spontaneous process is too slow or too inefficient a protein will evolve to speed it up or make it more efficient."

cell.) So the specifics can be distinct—we use this protein for memory, yeast for adapting to salt stress—but the core gears on which everything turns are preserved across all eukaryotes even if they look completely different. From an outsider's perspective, all eukaryotic cells are basically the same—bacteria and archaea would certainly say so.

What this means is that eukaryotic cells were nearly done evolving *before* all these various branches of eukaryotes came apart—so pretty early into our domain's tenure on the planet. At this stage, there was no such thing as an animal or plant—there were only single-celled microorganisms swimming in water. To be sure, eukaryotes were the largest and most complex creatures among them, compared to all the bacteria and archaea. In hindsight, we understand that these ancestral eukaryotes stood at the precipice of exploding complexity. They were soon to give rise to all the glorious and now-familiar branches of our domain, from ferns to mollusks to portobello mushrooms. These early single-celled eukaryotes had already developed molecules that later would be used by human neurons to store information, by fungi to break down starch, by plants to absorb water. But they had no way of knowing that yet. Even if they could think as well as we do, they would not have imagined the possibilities that evolution held for them. There wasn't anything more complex than single cells, and once the cells stopped getting more complex, what else was there to do? Had we interviewed those early eukaryotes around that time, they would no doubt argue that evolution is something that happened to them in the past and was largely over due to technological progress.

But boy, would they have been wrong.

How did eukaryotes break through their evolutionary stalemate? When it seemed that their cells couldn't get any more complex than they already were, evolution of eukaryotes changed course, like a river flooding over a dam. Instead of making their cells more complex, eukaryotes now focused on *doing more complex things with multiple cells.*

This summarizes two separate inventions that would become eukaryotic trademarks: sexual reproduction and multicellular organisms. On the surface, these textbook features of our body might seem like dull details—maybe they are important for how we function mechanically but hardly meaningful in a philosophical sense. Yet seen from the vantage point of early eukaryotes, sexual reproduction and multicellularity represent a complete reconsideration of what it means to be alive—a rewriting of rules, a biological New Testament of sorts. Its echo still reverberates in our lives, from our relationships with our ancestors to our relationships with our desires.

Sex Is for Change

The term "sexual reproduction" is very confusing because at its essence, the process is neither about reproduction nor about sexes. It is about making each new organism unique.

When bacteria reproduce, they clone themselves—in other words, their children are exact copies of their parents, save for an occasional mutation. But when humans reproduce, each child is totally different. This is because we don't just copy our parents' genes—we randomly pluck 50 percent of our genes from one parent and the remaining 50 percent from another parent. This randomness built into sexual reproduction creates new combinations of genes every time a new human is made and by doing so gives us the gift of individuality.

Reproduction is all about multiplying in number as efficiently as possible, and that's what bacteria do. In order to multiply cells, you need to multiply genes—DNA. Bacteria have streamlined the process: they have all their DNA wrapped into a single, neat circular molecule—simple and efficient.

By contrast, eukaryotic genomes—including the human genome—are a total mess. First of all, our cells (except our sex cells) actually contain not one, but two sets of genes, each originating from one of the parents. The sets are not just two copies, but rather two distinct versions of each other. So, for each given gene, each cell is simultaneously using two of its versions (also known as alleles). One allele may be slightly better, or worse, or just different from another, even though they are meant to do the same thing. This seems complicated: why these two redundant sets instead of just one gene for each function?

What's more, whereas bacteria keep all their genes tied up in a single large molecule that can be replicated in one go, eukaryotic cells carry dozens of separate chunks of DNA known as chromosomes. Because there are two sets of genes, there are also two sets of chromosomes: each one has a pair. Humans have twenty-three pairs, so forty-six chromosomes in total. Dogs, for example, have even more: thirty-nine pairs, for a total of seventy-eight. All these chromosomes are numbered and accounted for: we know, for example, that the human gene that controls alcohol breakdown sits on chromosome 4, whereas genes in charge of digesting proteins are located on chromosome 11. Since there are two chromosomes 4 and two chromosomes 11, there are two versions of each of the genes.

Why pairs, and why so many? It would seem that replicating one molecule is much more reliable than replicating forty-six or seventy-eight of them. Imagine packing your child a school lunch with seventy-eight menu items and not messing up a single one. If it seems like a recipe for disaster, it is. These endless chromosomes constantly end up in the wrong places; they break in half; they duplicate; sometimes one is lost; sometimes an extra one is gained. Every such aberration can be a cause for cancer or genetic disease if the aberrant cell is not identified and discarded. For example, Down syndrome occurs when people accidentally end up with three, rather than two, chromosomes 21.

So why not streamline reproduction, bacteria style, by combining this pile of chromosomes into one? Because reproduction is not the point. The point is mixing genes, and you need to have genes on multiple separate physical pieces to mix them.

This is how it works. When your parents manufactured the sex cells that would become you, they turned their forty-six chromosomes into twenty-three by randomly removing one chromosome from each pair. Which chromosome was removed from each pair in each sex cell of each parent was a roll of the dice, so each sex cell ended up with a unique set of twenty-three. Those unique sex cells of your parents are the source of your uniqueness. When the two sex cells united, the double set of forty-six chromosomes was restored but now with a novel combination of gene versions plucked from two parents. This double set was then inherited by all the cells of your body and only when it comes time to manufacture your own sex cells does it once again get randomly reduced to twenty-three.

In Spanish-speaking cultures, surnames are traditionally inherited in a similar way. For example, a man named Pablo might have two surnames, say Garcia Sanchez. There's the paternal half and the maternal half, and the parents themselves have two surnames (for example, Pablo's mom's name could be Maria Sanchez Campos, and his dad's, Diego Garcia Mendez). Just as parents lose 50 percent of their genes when passing them on to a child, Spanish-speaking parents lose 50 percent of their surnames. Historically, this system was heavily male biased—the first surname was always paternal, and it was also the one to be passed on. These days, parents can choose to give their children any of their surnames in any order, which is closer to how chromosomes are passed on.

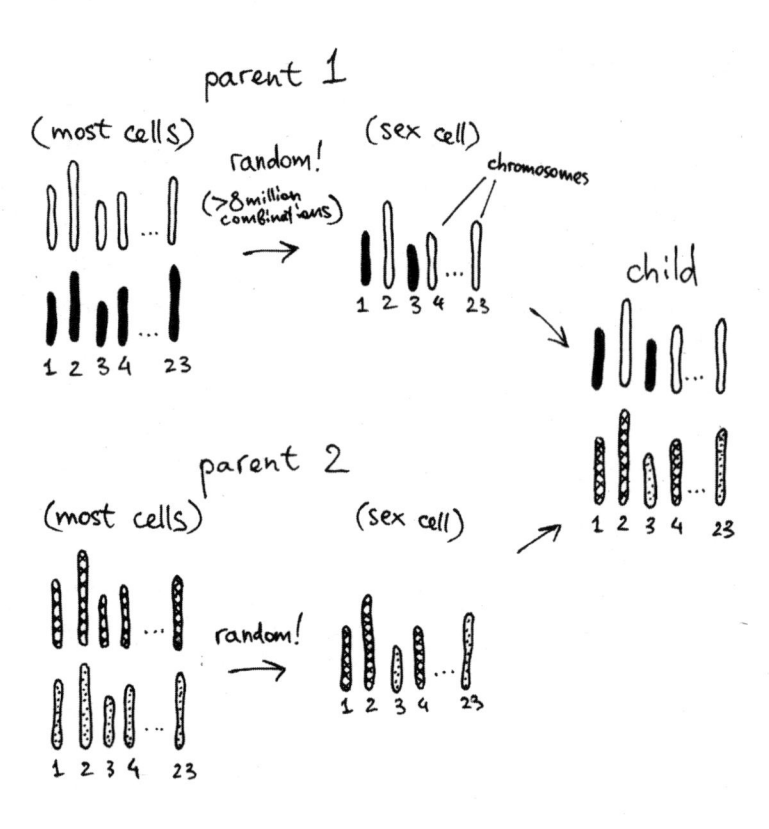

So we need this many chromosomes so we can mix them during sexual reproduction and create new combinations. If we had one chromosome per sex cell, there would be nothing to mix. No matter how you shook the "double" cell and how you then split it back into "single" cells, there would be nothing new: you would always return to where you started, like trying to make a cocktail out of two bricks.

The very arrangement of our chromosomes, in other words, presupposes that they will be mixed in each generation and prioritizes this mixing over the efficiency and even safety of reproduction. The more mixing in each generation, the more distinct each organism is from its kin. We eukaryotes seem to crave individuality even more than survival.

Breaking the Rules

Had bacteria been shown the blueprints of sexual reproduction at the dawn of the eukaryotic domain, they would no doubt be completely scandalized.

First, there's the reckless attitude toward DNA replication just discussed: a tangle of chromosomes instead of a single one. What could be more important than preserving the integrity of your genome, the repository of all knowledge accumulated during billions of years of evolution, bacteria would surely implore their strange, scary, overgrown eukaryotic neighbors. Surely it is a bad idea to keep the sacred scroll of ancient genes divided among dozens of sticky notes?

Second, bacteria would argue that losing 50 percent of genes in each generation is madness. Say you acquired a beneficial mutation—maybe a single error in one of your genes was so lucky that it gave you wings. But you have two versions of this gene, and only one has mutated. That means that each of your children will have only a fifty-fifty chance of inheriting wings. On the other hand, when a bacterium develops a beneficial mutation, all of its offspring get to use it without uncertainty.

Third, bacteria would say that having two sets of genes is bad for evolution. Let's imagine you again acquired a random mutation, but this time it's a dangerous one—maybe it disabled a gene responsible for detecting viruses. But again, you have two versions of this gene, and only one has mutated. That means you have a backup—even though one of the versions is broken, the second one still works, and you can still detect viruses. You therefore survive without even realizing anything is wrong with your genes. You then get a chance to pass on 50 percent of your genes to your children, which includes the mutated virus-detection gene. So instead of getting eliminated by selection, your mutated gene can continue covertly traveling through generations. Bacteria would say this is sloppy—bad genes must get weeded out! Instead, the mutation can continue getting passed down, undetected, until in one generation, by random chance, it matches it up with another chromosome that happens to have a mutation in the same gene. In this case, both versions of the gene would be broken, and there would be no working virus-detection gene to compensate. Suddenly, a sick child unable to detect viruses would be born to parents who seem completely healthy. This situation is particularly common if parents are closely related to each other, since their genes are already quite similar and likely have hidden mutations in similar places.

From a bacterial perspective, incest is the only good thing about sexual reproduction: at least it tends to expose hidden mutations, which stops them from getting passed on further.

But our perspective is different from the bacterial perspective precisely because we care more deeply about individuals than we do about the lineage. Just as treating bacteria with antibiotics kills most of them but strengthens the strain, incest is good for "purifying the bloodline," which was well known to Incan emperors, as well as Targaryen dragon lords in G. R. R. Martin's *Game of Thrones*. It is also standard practice for any animal or plant breeder attempting to develop a prized variety of dogs, horses, or tulips. But "purifying the bloodline" means a lot of "impurities" are rejected along the way: children unfortunate enough to be born with genetic diseases. It is no coincidence that incest, in most human societies, is a strict taboo—one of the clearest examples when consistent cultural norms grow around an underlying biological reality.

For bacteria, the only life that matters is the life of a lineage. Organisms are its disposable vehicles, and there are no other priorities than the ones set by the lineage. But eukaryotes unexpectedly elevate the disposable vehicles to a new status in which their individual priorities start to matter too.

Yes, our numerous chromosomes make cell division tricky—but it allows us a greater range of individuality. Yes, we lose 50 percent of genes each generation—but those that remain might create such a successful individual combination that this would compensate for the losses. Yes, we all carry a burden of dangerous mutations hidden in our genomes—but having a backup version of every gene protects against their potentially deadly effects.

All in all, eukaryotes set kin aside for the benefit of the individual. If our goals were to multiply ourselves as efficiently and faithfully as possible, we needn't steer away from the classic, bacterial-type asexual reproduction. But our goal, instead, is to maximize change, to increase variation among ourselves, and to bet that this will create enough new opportunities for success.

There's something else about sexual reproduction that bacteria would not be able to understand because of their collectivist nature. Sex not only gives us eukaryotes more randomness during reproduction; it also gives us *choice*. Despite using evolution to solve most of their problems, bacteria have no control over this evolution—they rely completely on the forces of nature to supply them with new mutations and pick the best ones. Eukaryotes, for the first time in the history of the world, get to have a say in what happens with their genes in the future. We can decide with whom we combine and reshuffle our genes. When a female bird listens to a male's song or when humans are swiping left or right on a dating app, that's basically what's happening: an organism evaluates potential future directions for the genes it carries. Yes, reshuffling chromosomes and discarding half of them in each generation might destroy any well-planned genetic arrangements. And yet, being able to at least partially design your future offspring based on the looks of your potential partner is a huge new advantage that is completely unknown in the bacterial world.

The Egg and the Ego

We've established that sexual reproduction is not about reproduction at all, but rather about variation and choice. What about sexes?

What exactly are sexes? When I ask students to define what sets the male and female sexes apart, the bio-savvy folks bring up X and Y chromosomes, and the rest mostly focus on the anatomy. The correct answer is typically a distant third.

Anything anatomical is too specific to humans, as is carrying offspring, parental care, and hormonal differences. If poplars or, say, single-celled algae have boys and girls, clearly there has to be a more general definition of what those concepts mean.

What about the X and Y chromosomes? Among our twenty-three pairs of chromosomes, we have this one pair, called the sex pair, which comes in two shapes, one known as X and another as Y. All eggs carry an X chromosome, whereas sperm can have either an X or a Y, and so when the two fuse, the resulting double set of chromosomes can be either XX or XY. In the first case, the embryo develops into a girl and in the second, a boy. But in other species—birds, for example—it's exactly the other way around.* In yet others—turtles, for instance—the sex chromosomes don't matter at all, and sex is determined by the temperature of the sand in which eggs are laid. So X and Y chromosomes *determine* sex—sometimes, not always—but they don't *define* what sex is.

What actually defines males and females is simply the size of their sex cells: male is the sex that makes the smaller one, and females the sex that makes the larger one. This is what all these various mechanisms determine: which organism

---

*To avoid confusion, bird sex chromosomes have different names: Z and W.

gets to make the sperm and which one the egg. This distinction may seem trivial at first sight, but it is actually the crux of all intersexual relationships in nature.

If the sex cells are the same size, then there's no distinction between male and female. This is called isogamy, and it was probably the original form of sexual reproduction. It still exists among some microorganisms—so, oddly enough, there can be sexual reproduction without sexes.[4] But by and large, modern eukaryotes have modified the strategy, making one sex cell small and the other large. This version of sexual reproduction is called anisogamy.

With isogamy, all sex cells are the same. So they all have to swim around looking for a partner. They also need to carry with them at least some nutrients to allow the future offspring to function. This is laborious and wasteful: the majority of cells will not find partners and so will perish with all their nutrients. By making sex cells different from each other, anisogamy allows one of them to focus on the resources and the other on the search, making the process more effective. A species reproducing in this way mass-produces millions of small, cheap spermatozoa, stripped-down genetic capsules propelled by a beating tail, and sets them loose to look for a small number of large, expensive, individually crafted eggs, which now don't have to move at all and can be packed full of nutrients without any restrictions.

(By the way, here's the solution to the chicken-and-egg problem: birds will not be around for maybe another billion years, but eggs—here they are.)

Males and females appear the moment one sex cell becomes larger than the other. Once the symmetry of sexes is broken, evolution keeps pushing their specialized roles further and further apart. The resulting differences are usually dramatic: there could be millions of tiny whizzing spermatozoa per one large, immobile egg. This cellular asymmetry in turn makes the two sexes approach sexual reproduction differently. Nutrients accumulated in the female egg are the primordial form of maternal care, and everything that follows is an elaboration of the same essence. The female sex by definition aims for expensive sex cells, and its priority is therefore to provide each one with nourishment and protection. The male sex, by contrast, aims for cheap sex cells and thus prioritizes their quantity and ability to seek out partners. Females tend to focus on the offspring, the males on looking for the females. Since there are way fewer eggs than there are spermatozoa, males also tend to face much stronger competition from members of their own sex and as a result often become more aggressive in the course of evolution. It all snowballs from that initial size difference between sex cells.

I hope it goes without saying that the relationships between the human sexes are not reducible to sperm and eggs. In the almost unimaginable time that has passed since the invention of anisogamy, entire worlds of genetics and culture have been layered on top of this single-celled logic of sexual reproduction, complicating and confusing everything to an unrecognizable extent. And yet it is equally clear that the origins of everything male and everything female in living nature are to be found here, in the merger between two ancient cells of different sizes.

The advent of the egg has yet another consequence besides the asymmetry of sex roles. What this giant new mother cell offers the embryo is more than just nutrients. It is also *time*. The life of a bacterium starts the instant its mother cell divides into two: the daughters must immediately start hustling for their own survival, feeding, and reproduction. But a giant cell packed with nutrients slowly accumulated by the previous generation can afford to sit around for a while. You could say that the primordial form of motherhood—nutrients in an egg—begat the primordial form of childhood. Now, a freshly created eukaryote did not have to start living immediately, but could first have a chance to grow, mature, develop, and with this luxury of time and resources, achieve a higher state of complexity.

It will not be long before not one, but multiple branches of eukaryotes—animals, plants, fungi, and brown algae to name the most notable ones—stumble upon just such a new state of complexity and transform living nature into the world we know today. Eukaryotes are about to invent multicellularity.

Queen and Commune

From human height, all ants seem more or less alike. They might look slightly different in color and size, but no more different than, say, makes of cars. But if we got down to their level, the various species of ants would seem as distinct to us as cats and rhinos. Some ants grow fungi in underground gardens and herd livestock (aphids). Some are tiny ambush predators that trap larger insects in holes of specially constructed fake tree bark. Others make elaborate homes out of leaves, clinging to one another during construction and using their own larvae as glue sticks. Yet others are terrifying murder machines, constantly fighting in clans within their own species and wiping out entire forests in the process.

What unites all these diverse creatures is their social structure, called "eusociality."* It is also found in bees, termites, and naked mole rats (not a joke). All the astonishing things that ants can do are carried out by a vast army of workers that selflessly labor and fight for the benefit of a fertile queen, their mother. United in their commitment to a single cause, worker ants can tackle much more sophisticated tasks than a crowd of self-interested insects such as cockroaches.

There is a very specific reason for this unity and selflessness: worker ants cannot reproduce—and never did. This means that a worker ant had never evolved by itself. All its genes are passed to it from the queen, its mother, which has been reproducing and therefore evolving. Because of this, the worker is essentially an extension of the queen's organism, a sort of remote-controlled

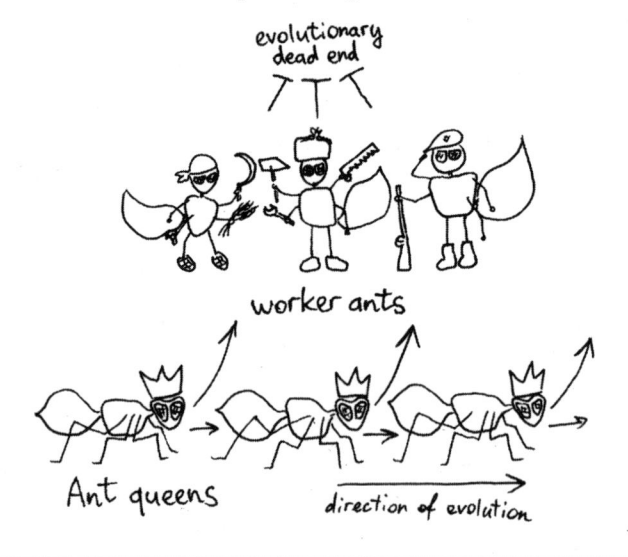

---

*To avoid steering too far from our subject, I must skip a lot of details that are extraordinarily interesting but require more explanation—for those, there is no better source than E. O. Wilson and his books on sociobiology and eusociality.

organ. It wants to build, forage, or fight for the queen because it is the queen who evolved the worker. It is she who provides the genetic instructions stating exactly what the worker should be doing. This is propaganda on another level. Ant colonies are sometimes even called *superorganisms*.

The relationship between the reproducing queens and the nonreproducing workers can be visualized as an arrow with offshoots. Generation after generation, the queen makes new queens, who go on to make new queens, and so on, but in addition each queen makes lots of workers, who make no one. Their lives are a dead end. Workers are an offshoot on the arrow that connects past to future through successive generations of queens.

This arrow is actually the main reason I am discussing ants. It is a useful metaphor to understand how our bodies function. If we now look at the diagram with ants and strip away the queens and the workers, we will see something else: the germ line and the soma, two components of a multicellular organism.

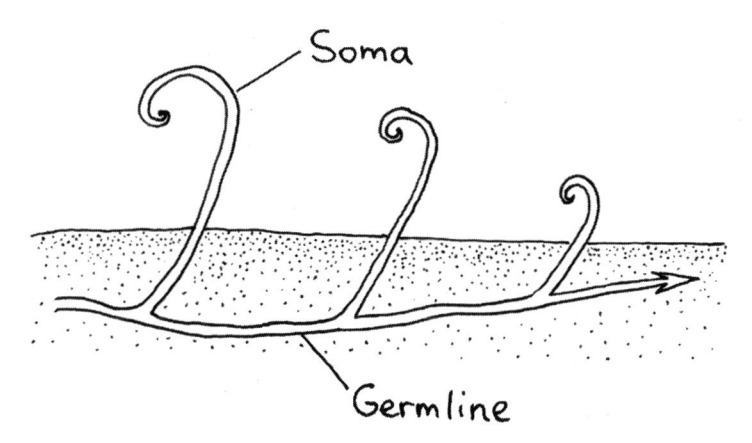

Just as an ant colony is more than a crowd of insects, a multicellular organism is more than a lot of cells. It is a lot of cells operating as one—and this unity, as in ant societies, is achieved by denying the majority of participants their reproductive rights. Most cells in our body undergo only a limited number of cell divisions, then stop, and only occasionally get replaced through new cell division. When the organism dies, they all perish together. In this, most cells in our body are like worker ants. Our organism consists almost entirely of these cells whose genes, strictly speaking, are destined to die. Skin cells, muscle cells, and brain cells all contain DNA, just like sperm and eggs, but skin DNA and brain DNA will never go beyond the confines of our organism and therefore will never make new humans. So most cells in our body are facing an evolutionary dead end. The only cells that have a chance to leave a genetic mark in history are members of a special, privileged group equivalent to queen ants: sex cells and their precursors in the testes and ovaries.

These two "castes" of the organism's cells—the mortal majority and the immortal minority—are called the soma (Greek for "body") and the germ line.

If you tried to explain multicellularity to bacteria—already flabbergasted by the notion of sexual reproduction—you would certainly confuse them even further. You'd have to tell them that our human *germ line* is more or less like the bacterial *strain*—a continuous line connecting generations of cells in time. But you would be at loss to explain what the soma is: this additional pool of subjugated cells that help the main "germ line strain" proliferate and then die off without continuing their own strains. What would be most alien to bacteria is the fact that *we think of ourselves as those mortal, subjugated cells*. For us humans, the soma is primary, and the germ line secondary. Our body, our brain, our consciousness—everything we call "I"—is part of the soma, the evolutionary dead end produced as an accessory to the germ line. Bacteria would have been astonished at our evolutionary pointlessness.

Indeed, it all sounds dreadful: subjugation of the mortal soma by the immortal germ line! Is a man really just a tool for his spermatozoon? Surprisingly, I would say the opposite is true: the soma–germ line split is the main source of our personal freedom.

Had we been unified with our germ lines, we would not have had any chance, ability, or motivation to ever go against our genes, just like bacteria, who wouldn't have even understood the concept of "going against their genes." In general, yes, the soma follows the genetic instructions handed to it by the germ line, because if it doesn't, the germ line dies out, and the soma disappears. But the germ line simply cannot control everything the soma does. As evolution makes the somas more complicated, the germ lines have to become content with only giving them general guidelines rather than specific instructions for every situation: don't die, eat food, have sex. The number of parts in the human body vastly exceeds the number of genes in the human genome. And yet all the instructions for all these parts must somehow fit into a single cell passed into the next generation—an egg or a sperm. This means that the more complex a multicellular organism gets, the less specific its genetic instructions become, and the more space for personal freedom it is allowed. Instead of blindly relying on genes, the soma must figure out its own problems, which it does both by internal cell-to-cell communication and by surveying the external world. In animals, this drive toward personal freedom culminates in the invention of a special organ that continually learns and applies new information that is gained from experience and not featured in the genes. This organ is called the brain.

Just as sexual reproduction allows us to be distinct from our parents, multicellularity allows us to think for ourselves. The germ line may have created the soma for its own benefit, but by giving it partial autonomy, it opened the door for the birth of the individual mind.

Emergence as a Credo

For all their groupthink, bacteria, ever the collectivists, always remain individual cells. Eukaryotes, having embraced individualism as a life principle, can't seem to keep themselves from banding together into various hypercells. A cell with two sets of chromosomes is the first such hypercell. The multicellular organism is the second one. A superorganism of ants is a hypercell on yet another level. Paradoxically, having stepped into our eukaryotic empire of predators and egoists, we became a lot more interested in one another.

There's a broader principle at play here, not just cells sticking together, but new things originating from combinations of old things. Life is a layering of levels, each of which is built from the components of the previous one but not reducible to them. A multicellular organism is more than a lot of cells. A cell is more than a handful of chromosomes and proteins. A molecule is more than a scaffold of atoms.

This phenomenon has many names: synergy, holism, systemic effect. I use the word *emergence* because this is how it was introduced to me by Professor Andrei Granovich in his first lecture on invertebrate zoology on day one of my freshman year of college. I have to confess that in that specific moment there were lots of other more important things going on, so the organizational principles of living matter were low on my priorities. I wrote it all down mindlessly and forgot about emergence until the end of the semester. But I have a distinct memory of the moment when, preparing for the final exam, I reread the first page of my notes and suddenly *got it*. Everything. How life arises from nonlife, how it finds new ways of being, how its complexity increases in the process, and how it looks as a result. *Emergence!* To me, higher education exists for moments like this. I am still grateful to Professor Granovich for giving me this concept as well as a general passion for the hidden meanings of life.

Emergence is a simple idea: the whole is greater than the sum of its parts. A molecule is not just a few atoms but also their specific configuration. A sentence is not just a set of words; it is also the meaning that arises (or *emerges*) from their interrelationships. A melody is not just a few Cs and a few Ds; it is also the harmony produced by their mutual arrangement. In short, an emergent system has properties that are not reducible to the properties of its components.

But in the case of living nature with all its perpetual evolutionary motion, emergence is more than just a convenient descriptor.

It is a long-term survival strategy, an approach to creating new things where old possibilities appear exhausted. It is this art of creating new levels out of old ones that eukaryotes have mastered so well. When it appeared that their cells could not get any more complex, they created new levels of complexity beyond cells and reconfigured the very essence of reproduction to sustain those new levels. The result was a cornucopia of new possibilities—the evolutionary roadblock became instead a launchpad. Almost everything we care about in living nature, from orchid petals to octopus tentacles, is built out of emergent combinations of very similar components. Emergence does not just create larger life-forms—it opens up new dimensions in the medium from which evolution can mold life.

And what of the seemingly stagnated evolution of our own species? Ethics and medicine may have cushioned humans against the raw power of natural selection—just as having two sets of chromosomes at one point cushioned our ancestors against dangerous mutations. But if history is any indication, evolution always finds a way. Maybe we won't grow wings or ear covers. Maybe our future evolution won't be about our individual bodies at all, as much as about things that we do together—our cultures, our societies, our languages. Maybe our evolution already continues on a new level of emergence, and we simply fail to recognize that it had never really stopped.

Part II.
Where We Came From

Chapter 5.
The Moving Kingdom

"There is no motion," said one bearded thinker.
Another walked in silence back and forth.

ALEXANDER PUSHKIN

My favorite part in any adventure novel or epic saga is seeing how multiple factions of characters pursue similar goals in different ways: Gryffindor and Slytherin in *Harry Potter*, Lannisters and Starks in *A Game of Thrones*, elves, hobbits, and orcs in *The Lord of the Rings*.

Nature is also like that except on a much grander scale than any saga.

In previous chapters, we saw that the origin story of humans truly begins with the birth of eukaryotes. This moment and the great inventions that followed it—sex and multicellularity—paved the way for life as we know it today: not just a soup of microbes, but an absurd bacchanalia of colors, shapes, sounds.

But like a young, burgeoning empire, the nascent domain of eukaryotes did not remain unified for very long. Soon after its arrival on the historical scene, it fragmented, splitting into countless kingdoms and fiefdoms. Each of them pursued the same goal, survival, but each in their own unique way.

Some of these groups of creatures remain relatively obscure to this day—for example, slime molds known to display a rudimentary form of intelligence or trypanosomes that cause sleeping sickness. But some kingdoms—most notably

eukaryotes

slime molds   algae   plants   fungi   animals
trypanosomes   many others!

fungi, plants, and animals—have amassed great power in today's nature. Each of them prospered because it did something better than anyone else.

How plants distinguished themselves is more or less clear. The "green kingdom" stands on the bedrock of photosynthesis, even though plants did not invent it—bacteria did. Plants, however, took photosynthesis to a new level. They originally appropriated it from cyanobacteria by incorporating them into their cells as chloroplasts. But then, in true eukaryotic fashion, plants were not satisfied with the magic of photosynthesis as it was—they wanted more. They achieved that through multicellularity—assembling into gargantuan, stationary sun-harvesting stations that ended up sucking nearly all carbon dioxide out of the ancient atmosphere and turning it into living matter.[1] Today, the success of plants—the green paint over the Earth's continents—is visible even from the moon. Collectively, they outweigh the rest of the biosphere put together and not just by a little, but almost five to one.[2] Plants may not be able to claim any world-shattering biochemical breakthroughs comparable to the invention of photosynthesis by bacteria, but could any bacterium imagine a sequoia tree?

Another great kingdom of eukaryotes is fungi. Fungi, like animals and unlike plants, need food to survive. Where the life of plants revolves around photosynthesis, the life of fungi revolves around finding something to break down and consume. Fungi are extraordinarily good at digestion, thanks to an enormous variety of powerful enzymes that can decompose virtually any scrap of organic matter down to the last atom. They also use their wondrous chemical abilities to take advantage of other creatures: as we will see, throughout evolution fungi repeatedly recruit plants and algae to produce food for them, and they can even turn insects into zombies that spread their spores. (The reason this doesn't seem to happen to vertebrate animals—a scenario from the zombie apocalypse video game *The Last of Us*—appears to be the aversion of most fungi to our high body temperatures.*)

So plants are food factories, and fungi are mad scientists. But the essence of our own kingdom, the animal kingdom, is far from obvious. We are not the largest nor the most numerous nor the most long-lived. We cannot photosynthesize. We cannot digest nearly as many things as fungi can. There are organisms out there that are tougher, more widespread, and have more complicated cells. So what is it that makes us special? What do animals do that no one else does as well?

Maybe the word itself can offer some clues. Which qualities do we humans put into this concept, *animal*? The English word *animal* is derived from the Latin *anima*, a word meaning both "breath" and "soul," and ultimately the Proto-Indo-European *henh*, "to breathe, blow." So, then, an animal is a *creature that breathes*. This one won't do: as we saw in chapter 3, breathing, like photosynthesis, is a bacterial invention that was appropriated by all eukaryotes, not just animals. Although it wouldn't have been obvious to Proto-Indo-Europeans, plants and fungi can also breathe, so it can't be the trademark of animals.

The Russian for "animal," *zhivotnoie*, is also not much help. It descends from the old Slavic *zhivot*, which today means "stomach," so "animals" sounds a bit like "stomach creatures." The word *zhivot*, however, is related to both the Latin *vita* and the Greek *biota*, both of which mean "life," and *zhivot* also used to mean "life" in the past, so the real meaning of *zhivotnoie* is *living creature*. This

---

*The plot of The Last of Us, which is also a TV show, is based on the premise that fungi break this temperature barrier, sort of like viruses break the species barrier, and go on to infect the entire humankind. This is unlikely in reality because our brains are very different from the brains of insects, and you can't so precisely control both using the same techniques. To alter our behavior in the way they alter the behavior of insects, these fungi probably would have to evolve with us for millions of years, all this time infecting us in more prosaic, non-mind-controlling ways, just like other germs, and only gradually accruing more ingenious ways to alter our brain chemistry. By contrast, for example, a bat virus, which normally cannot infect human cells, only needs to tweak a couple proteins to start doing that. So a virus spillover from bats requires exponentially less change than a spillover of mind-controlling fungi from insects.

is an even worse way to define an animal than *creature that breathes*—living creatures must also include plants, fungi, and many others.*

What about other languages? From all the examples that I know, the best one—the one that, in my opinion, most accurately reflects the real essence of the animal kingdom—comes from Chinese. In Mandarin Chinese, an animal is known as *dòngwù*, which is written as two characters: *moving thing*.

Animals move. Think about it. In the absence of obviously identifiable features like eyes or legs, that's how we know if something is an animal: active motion. When we see a gooseneck barnacle on a beachside rock, we might stare at it in doubt for a while, unable to decide if it's a specialized plant or maybe an unusual fungus, but when a wave hits the rock and the barnacle opens up its shell, extending whiskers gratefully into the swirling sea foam, we delight in recognizing it as one of us—animals. Rare examples of moving plants—such as the sensitive mimosa, which folds its leaves in response to a crawling insect—are wonderful precisely because they remind us of ourselves. (Conversely, rare instances in which animals appear still—like sponges or anemones—elicit, at least in some of us, a strong desire for poking, as if to verify their allegiance.) The "breathing" associated with the soul in the Latin word *anima* really refers to *visible breathing*—that is, breathing with motion, such as the rising and falling of the chest—rather than to respiration as a biological process.

Animals and fungi—two neighboring kingdoms within the eukaryotic domain—share their need for food but pursue nourishment in fundamentally different ways. Fungi don't move—they grow, slowly and steadily, relying on complex chemistry to wrangle the environment into submission. We animals may not have such a level of molecular control over the world, but what we have instead is fast motion, which allows even a small animal like a bee to harvest food from a vastly greater territory than even the largest fungus in the world (which, to date, happens to be a contiguous specimen of *Armillaria ostoyae* in northeastern Oregon—possibly the largest organism on Earth, weighing about as much as five thousand African elephants and colloquially known as "Humongous Fungus.")

---

*"Stomach creatures" actually works a lot better but unfortunately is also imperfect, leaving out some important members of the kingdom.

I think that subconsciously we all understand that motion is the central animal ability, even if we rarely articulate it. Consider the word *animate*: even though it comes from the same root *anima* and so technically means "to breathe life into something," the meaning that it really conveys to us all is "to put in motion." An animated movie is a movie made of moving pictures. An animatronic doll is a moving dinosaur on stilts. At Turkish resorts—I visited one with my parents as a teenager—an *animatör* is a guy who makes people dance.

Every interaction we animals have with the outside world boils down to motion. Right now, there are lots of complicated things happening inside my organism, but the only consequences they have on the environment is the motion of my fingers across the keyboard and the motion of the coffee from the cup into my mouth. All work—really, all *behavior*—is physical movement, whether it is the movement of boxes, the movement of a steering wheel, or the movement of air by vocal cords. To influence the world, an animal must move something.

Even individual animal cells reveal a bet on motion above all else. One cellular structure that almost all other creatures in the world have but animals lack is a cell wall—rigid armor surrounding the thin, flexible cell membrane. Bacteria, archaea, plants, fungi, and algae all have different kinds of cell walls, and the sheer variety makes it clear that a cell wall is an attractive evolutionary idea to which many groups have repeatedly turned. But not animals. Why not? Because cell walls, as useful as they are in protecting cells from damage, would stand in the way of motion. With cell walls, an animal body would be better defended, but its skin and muscles would have all the flexibility of celery. As they are, animals are quite vulnerable to physical damage compared to plants—stomp on grass and it will still grow; step on a caterpillar and it will probably be the end of it—but jettisoning cell walls gives animals freedom of motion no plant could dream of. An animal is made practically out of foam, soft cellular bubbles that bend, crawl, and contract in all dimensions. Wherever you push, something pops, rips, or breaks—but such is the price we animals pay for running fast and biting hard.

Plants have made different gambles in their evolutionary history and so face a different set of problems. Because of photosynthesis, plants don't need to look for food and so can afford to stand around in one place. This is not to say that plants don't move at all. Watch a time-lapse video of plants growing in a newly cleared patch of forest and see violent jousting with hooks and spiked

Animal cell   *Plant cell*

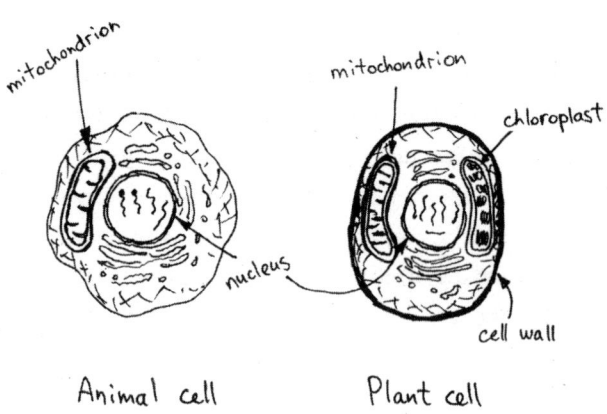

clubs to secure a place under the sun—quaint and bloodless it may seem to us, but on the scale of days and weeks, no less ferocious than animals fighting for food. And yet plants could not possibly match the mobility of animals. The fastest known plant motion—the shutting of the Venus flytrap—takes about 100 milliseconds,[3] which is very animal-like but still almost forty times as long as it takes a mantis shrimp to strike its prey.[4] No plant can perform such fast motion continuously, and none can lift its entire body from the ground and replant itself into a new location, as an animal migrates from place to place.

There is a curious twist to this motionlessness. Plants, as a matter of fact, do require long-distance motion—for dispersal. If you can't move, you must somehow ensure that your progeny doesn't end up in the same place as you—otherwise, you'd be competing with your own kin for the rest of your life. This dispersion issue causes plants a great deal of trouble. Some of them, like ferns and pines, rely on the movement of wind or water to spread their spores or pollen. But the most successful group of today's plants—angiosperms, or flowering plants—owes its dominance to an ingenious strategy: partnership with animals. Flowering plants are relatively new: they exploded in diversity only within the current geological era. Today, however, they form the bulk of the biomass in a jungle or any broad-leaved forest and provide almost the entire caloric intake for humankind—all the fruits, vegetables, and grains, including those eaten by animals that humans eat later. Flowering plants are juggernauts of modern flora, and they can't get enough of us. Their taste, their colors, their smells are all designed to attract members of the animal kingdom for the purposes of dispersal, which can happen either through pollen in flowers or through seeds in fruits. The world of flowering plants is an assortment of whimsical devices targeting the brains of potential pollinators or consumers. Any edible berry, any beautiful flower, any aromatic nectar is an evolutionary investment by the plant kingdom into the animal kingdom. I imagine that in the old days, plant ancestors would have scoffed at such frivolities. Why would plants, these almost perfect creatures that possess the power of photosynthesis and depend on almost nothing in the material world, suck up to animals, freeloaders always looking to bite a chunk out of them? The answer is motion. No matter how much pollen a bee eats, it still does a better job of passing the remaining pollen from flower to flower than wind does.

So modern flora and fauna are more intertwined than they ever have been, owing to this recent alliance between two great kingdoms of eukaryotes, the green one and the moving one.

How did we animals distinguish ourselves in this particular animated way? Bizarrely, it is not a simple story of inventing a new superpower. Rather, it is about rediscovering an ability that has been lost. In a seemingly mythological sequence of events, the founders of the animal kingdom were initially banished from the ranks of "moving things"—only to fight their way back to the top.

The Children of Sponge

When animals first appeared, motion was not, by any stretch, a new idea in living nature. It had always existed in the single-celled world—as far back in the history of life on Earth as we can tell. Even bacteria and archaea can move, and more complicated single-celled creatures like ciliates can dart around in all directions, twisting and tumbling with great speed and agility as they search for food or a mate. They achieve that using specialized propulsion devices attached to their cells: cilia or flagella, which beat or rotate and so push the cell in various directions. Ciliates in particular are completely covered in cilia—hence the name—which is part of the reason their movement can be so sophisticated.

The problem is that this only works well for single cells. Multicellularity, on the other hand, weighs the body down, making cilia-based propulsion inefficient, like steering an aircraft carrier with canoe paddles. The largest organisms that propel themselves with cilia are mysterious jellyfish-like animals called ctenophores, and that's pretty much how their motion looks: a slow drift through the ocean's depths powered by tiny paddles of bundled cilia. Ctenophores are beautiful, ethereal creatures, producing a wonderful iridescent effect with the synchronized beating of their cilia paddles. They are very large for their mode of locomotion, but they still rarely get past a few centimeters in size and move a lot slower than real jellyfish, let alone a squid or a tuna.*[5]

So the more cells you stick together, the more difficult it gets for them to move as a unit. This is the issue that animal ancestors faced at the very dawn of the animal kingdom, when they were just one of several groups of eukaryotes experimenting with multicellularity. All animals are multicellular by definition, meaning that the border separating animals from nonanimals on the evolutionary tree is deliberately chosen in such a way so as to correspond to a transition from single-celled to multicellular organisms. This border is actually a perfect illustration of multicellularity weighing the body down. On one side of the border, we find choanoflagellates—single-celled creatures that are fast and confident swimmers, evoking swift microscopic tadpoles or fat sperm cells with collars around their necks. They are the closest single-celled, nonanimal thing that there is to an animal. On the other side, we find sponges—animals that look so still and motionless that it can be hard to think of them as animals at all, despite their multicellularity and large size. If the animal kingdom begins with stillness, how did we ever become the kingdom of motion?

The key to the answer is food—that one thing that plants have figured out for themselves, but we animals perpetually need. It is food that pushed us to become multicellular and lose our primordial ability to move; it is food that will push us to rediscover motion at a scale never seen before.

Comparing choanoflagellates to sponges is our best bet at reconstructing the first moments in the history of the animal kingdom.[6] Both groups of organisms sustain themselves chiefly on a diet of bacteria. A choanoflagellate captures these bacteria by beating its flagellum and so creating water flow through its collar, which consists of long protrusions of the cell membrane arranged tightly in a circle around the flagellum. Bacteria are captured in these protrusions like shrimp in the bristles of a whale's baleen. Sponges also feed chiefly on

---

*Ctenophores are mysterious because no one is quite sure how they fit into the tree of life. Because of their general jellyfish-like demeanor, they were historically grouped with cnidarians, which includes jellyfish and corals. But some scientists argue, based on genetic evidence, that ctenophores are in fact the oldest existing branch of animals, whose evolution has run parallel to the rest of the animal kingdom since its very beginning—even before sponges were in the picture. This is still an area of active debate.

Choanoflagellates

Choanoflagellates feed on bacteria trapped in the collar

water flow

Sponges

Sponges resemble stationary batteries of choanoflagellates

— (all other animals)

Metazoa (animal kingdom)

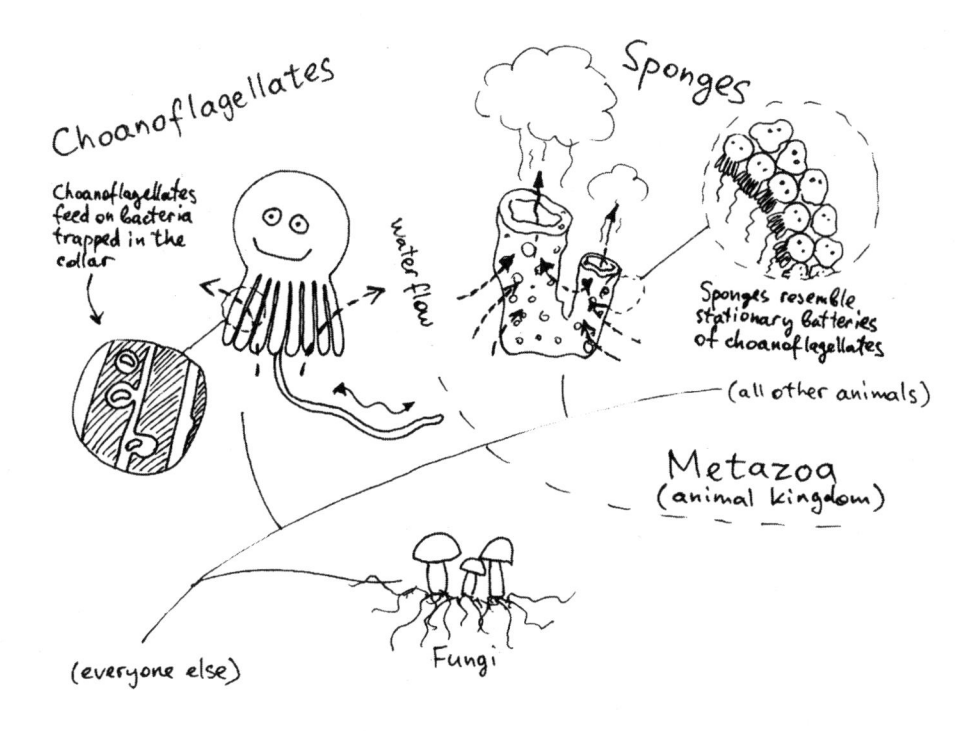

(everyone else)

Fungi

bacteria, which they capture in the collars of cells that look almost exactly like choanoflagellates, except they form a continuous layer that lines the inside of the sponge. The collared cells share their bacterial catch among each other and also pass it to supplementary cells that are not directly engaged in capturing food. It's as if a sponge were a bacterium-harvesting factory, whereas choanoflagellates were individual artisans doing the same thing by themselves. The line separating the two is not as sharp as it might seem—many sponges only loosely adhere to their multicellularity and don't mind being pressed through a sieve, reforming back into a sponge after a while.

The reason that sponge cells stick to the factory rather than harvesting food on their own is improved hydrodynamics. The total flow of water through a layer of sponge cells is greater than the sum of individual currents that can be generated by each cell.[7] Sponges can achieve truly impressive power: look up "sponge pumping" and observe the fountains of water discharged from these seemingly motionless creatures—the flow can be made visible by dyeing the water around the sponge. Interestingly, choanoflagellates themselves occasion-

The Moving Kingdom **83**

ally do something similar on a smaller scale: they band into spheres of tens or hundreds of cells, like tadpoles with their heads put together.[8]

In other words, the kingdom of animals was founded upon not the invention of body motion, but rather its rejection. The ability of single cells to move was traded instead for collective labor that yielded more food per capita. This collective labor, though, still involved motion—just not the motion of the body, but rather the motion of water through the body. That's what *hydrodynamics* means: water motion.

Sponges became multicellular to harvest more food, and that made them settle down on the seafloor. This, in turn, created a new problem—dispersal. Just like plants have to deal with dispersing their seeds while mostly remaining stationary, sponges also needed a way to ensure that they could spread from one place to another: however still you are, something must move somewhere at least once per generation. And so, after a stationary egg inside a stationary sponge is fertilized, it develops into a larva—a moving ball of cells that takes off and swims away from its stationary mother. The larva explores the world for a couple of days before settling onto the seafloor in some distant location and gradually assuming its large, calcified adult form.

Aside from their pumping abilities, adult sponges are admittedly not the most *animated* creatures in the world. But sponge *larvae* do look and behave like typical animals, despite their microscopic size. They consist of a few hundred cells and have a neat, symmetrical body shaped like a little cantaloupe. They are motile and energetic, although they don't have any muscles and can only swim using cilia. They react to the environment. Some even have an

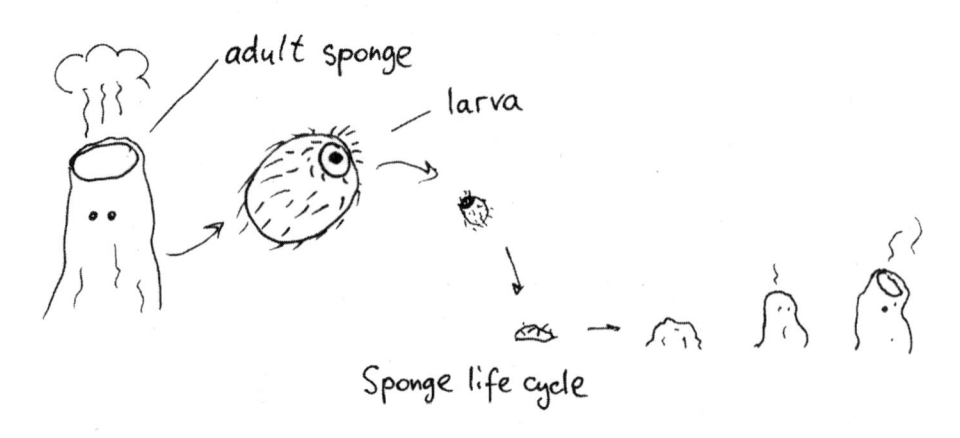

Sponge life cycle

84 One Hand Clapping

eye—a photosensitive spot reminiscent of a headlight, except technically it's in the back of the body, since the goal of the larva is to swim away from light.[9]

It makes sense that many evolutionists trace the descent of the remaining animal kingdom from these baby cantaloupes rather than actual adult sponges: all animals other than sponges themselves, they say, are variations of a sponge larva.[10] At some point, some sponge larvae decided they didn't want to become motionless stone chimneys and instead evolved into a separate species, taking off on their own evolutionary journey—one that would end up eclipsing every expression of nature that had existed to this point.

If you were an adult sponge at this time, though, you would surely think that your larvae becoming their own species and evolving into anything other than larvae is an extremely bad idea. (Imagine a human baby evolving into a new species of human, consisting only of babies, without ever learning to speak.) The whole reason for becoming multicellular in the first place, you would no doubt reason, was to create a large static body that could harvest abundant bacteria. It is this food harvesting that allows the larvae to even exist! (A sponge larva is not meant to live for long and usually doesn't have a way to feed itself—it survives on nutrients provided by its parents until it becomes an adult.) The body of a sponge larva may have motility, but it is pretty useless otherwise—or so you would have thought as you sullenly sifted through your bacterial chow in the ancient ocean. What life could these youths possibly envisage on their own?

What you would not have anticipated is that sponge larvae, these useless balls of moving cells, were about to discover a mode of feeding that would forever leave their ancestors in the dust.

The Communal Kitchen

There is a big difference between how sponges feed compared to other animals, including humans. The most notable distinction is the fact that we have a mouth.

In a sponge, each individual cell grabs its own bacteria, engulfs them, and digests them internally, using its own cellular enzymes. In our bodies, digestion is centralized. Communal enzymes break food down in the gut cavity, and the resulting nutrients are then absorbed and delivered to all the cells using a centralized distribution network, blood. It's as if sponges had every cell cook their own food in their own pots, whereas human cells ladled their food out of a big communal pot. What this means is that sponges are restricted to microscopic food—anything multicellular is generally out of their league, because no individual sponge cell can swallow a whole shrimp.*[11] We, however, can devour food in large chunks—including whole other animals—thanks to our centralized digestion.

To achieve this centralized digestion, an animal body needs to have two key features: an internal cavity where the communal enzymes can be contained—the gut—and a valve that would open and shut this cavity—the mouth. (If the cavity can't open, you can't get the food in, and if it can't shut, all the enzymes diffuse away.)

It's a pretty neat system, really. It is not a trivial task for a bunch of cells to figure out a way to wrap themselves around a massive piece of prey, secure it, and blast it with enzymes. It was achieved through a simple maneuver of folding the ball of cells—the sponge larva prototype of all other animals—inward, like a Ping-Pong ball pressed with a thumb. This fold gave rise to exactly the kind of body organization that could support centralized digestion. It created an internal pouch—a gut, or stomach, which at this stage is all one thing—and a mouth.

In fact, the folding of the Ping-Pong ball proved so consequential that we all continue doing it to this day as we develop in our mother's womb. At some point early in our embryonic development, every human resembles a sponge larva—a sphere of cells without discernible body parts. Then, at a critical stage called gastrulation ("stomachation"), this sphere of cells folds inward, just like our ancient ancestor once did. This is the preface to all the other events that happen during our embryonic development, and it was a preface to the evolution of all the other animals that would soon roam the planet.

Remember the origin of eukaryotes? What made our domain explode like a supernova was the ability of eukaryotic cells to swallow other cells whole, previously unknown in the world of bacteria and archaea. This gave eukaryotes access to unprecedented quantities of concentrated energy that they could seize from others.

---

*There are some exceptions: the harp sponge *Chondrocladia lyra*, discovered in 2012, can entrap and slowly digest larger prey, like small crustaceans, using special hooks between its calcified vertical spikes resembling a harp.

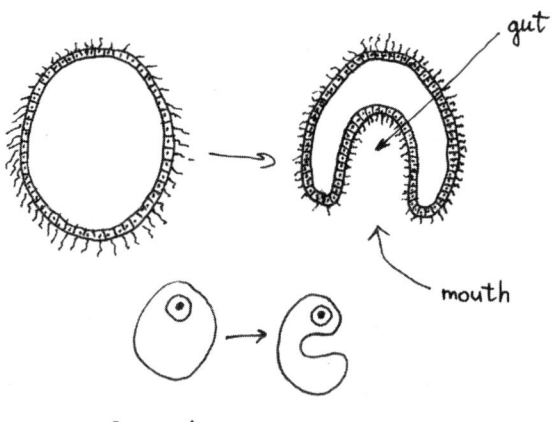

Sponge larva stomachates

By creating a gut with a mouth, the folded sponge larva achieved almost the exact same breakthrough, except at the multicellular level. It now had a cavity where not just bacteria, but large, multicellular organisms could be placed and digested whole—swelling the larva's energy budget and forever embarrassing its hard-working, bacteria-sifting parents.

In fact, we shouldn't call it a sponge larva any longer—the new animal was so different that we need to give it a new name. Sponges—whether adults or larvae—do not swallow each other. But *jellyfish* sure do.

A Gutsy Move

A jellyfish, or more scientifically a medusa, is actually one of the two forms of animals called cnidarians, which translates as "nettles animals," referring to their stinging abilities reminiscent of the nettles plant. The second cnidarian form, which complements the medusa, is called a polyp. Many cnidarians switch between these two forms in the progression of their life cycle: swimming, pulsating medusas can settle on the seafloor and turn into sessile, flowerlike polyps, and vice versa. Corals are made up of a multitude of tiny polyps. Hydra—a microscopic classroom favorite—is also a polyp but a singular one.

As a whole, these animals are a highly successful group. In the ocean, there are cnidarians everywhere, in large numbers, and in an astonishing variety. At first glance, there is nothing shared by creatures such as the Portuguese man-of-war with its translucent flotation bubble and blue beard of long killer polyps; the fierce anemone that hides clownfish in its bright tentacles; or the solemn brain coral that looks like a rock carved by an alien civilization. But if you strip any cnidarian body to its basic configuration, it's that same Ping-Pong ball pressed inward, except a jellyfish points the opening—its mouth—downward, and the polyp points it upward.

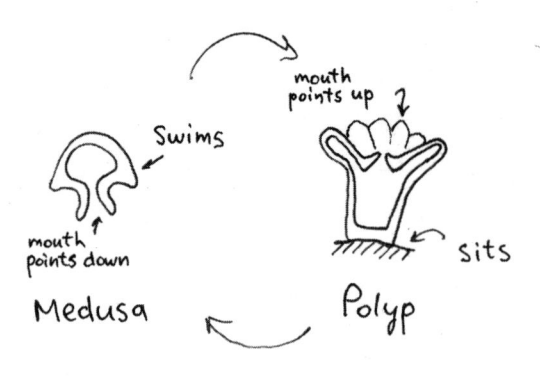

Evolutionarily speaking, polyps are actually the original form of cnidarians; jellyfish are a later spin on the same body shape.[12] So the renegade sponge larva that folded inward did not transform into a mobile jellyfish right away, but rather into a stationary polyp—not that different from the lifestyle of its sponge parents, who also remained stationary and presumably pretty upset about all this folding with which their larva was experimenting.

The polyp now could eat much bigger food than its predecessors—if the food happened to fall upon it.

But true eukaryotes are not satisfied with scraps. The new ability to digest large food was too attractive to remain motionless. Soon enough, the polyps started moving around, crawling and bending to find prey, and even peeling off, flipping upside down, turning into medusas, and using their mouths and guts to swim, grab, and steal energy from living, breathing creatures, even those that may even attempt to run away or fight back. This is what gave them the edge. If ancient eukaryotes prospered because they invented predation of cells by other cells, you could say that cnidarians pioneered predation of animals by other animals.

88 One Hand Clapping

So it is food, once again, that put the animal kingdom back in motion. And this time, it was not just mere beating of cilia. For the first time ever, animals propelled themselves with *muscles*.

Muscle cells are some of the most ingenious cells in our bodies, on par with neurons, with which they share many similarities. Like neurons, muscle cells can get electrically excited, but unlike neurons, this electrical charge causes them to physically contract with great force. Each muscle cell is packed with long strands of a protein called actin, and these strands are interlinked with bundles of a moving protein called myosin. As the cell receives the electrical signal, myosin proteins run toward each other, which slides the strands of actin toward a central point, and that in turn contracts the cell. It takes a lot of energy—but produces a lot more power than the beating of a cilium.

Although actin and myosin exist in other eukaryotes (for example, they are used to break the cell in two during cell division), their use for muscular movement was pioneered by cnidarians.[13] Polyps used contractions produced by these proteins to wiggle their tentacles and occasionally to move around from place to place to seek food or protect themselves from the elements. Medusas used the same contractions to turn their body into one large, rhythmically contracting propulsion device. At the time, this was all new.

There was something else that this jellyfish and polyp motion required: a nervous system. To produce a useful movement, such as a contraction of the body or a rhythmic pulse, many cells must quickly coordinate their activity in time. A nervous system, in its original cnidarian iteration, provides just that. Like cnidarian muscles, it is not yet clearly separated into a distinct organ but permeates the skin of polyps and jellyfish as a network of interconnected neurons.[14] These cells are electrically charged and can use this charge to rapidly pass a signal from one part of the body to another. Other cells in the animal body also have the ability to communicate, but they typically send their signals to each other in the form of chemicals such as hormones. Chemicals take a while to diffuse from one place to another, whereas electricity spreads very fast. This is what the nervous system originally evolved to do: faster signal transmission and so better coordination of polyp and medusa muscles. A few hundred million years from this point, this supplementary system of motion control would become the seat of the human mind.

But there is a long way to go before that. For now, the kingdom of animals has won its right to be called the kingdom of moving things and will forever adhere to this title.

Notice, however, that even the movement of jellyfish, which are much more mobile than polyps, is, to this day, severely limited compared to the majority of animals we know: *they can move in only one direction*. The invention of muscles and nerves allowed multicellular bodies to move faster, but it did not immediately give them the agile, flexible, adaptable movement that cilia and flagella give single cells.

This was about to change.

Left and Right

In the movie *Arrival*, based on a novella by Ted Chiang, humans establish contact with an alien civilization that turns out to have an entirely different conception of time—they don't recognize beginnings and endings, but rather think of all events as occurring together. In the story, the explanation given for this cultural difference is ultimately the shape of the aliens' bodies—rather than having a front, a back, a left, and a right as we do, the aliens are radially symmetric, like jellyfish, and so, the story goes, have never evolved to understand space and time in the same way as we have.

Although this is science fiction, it is indeed difficult for us humans to wrap our minds around a radially symmetric body, meaning symmetric around a central axis. It's not enough to imagine yourself with ten arms and ten legs hanging evenly on all sides of the body—in this fantasy, you'd still be able to control all these limbs independently, whereas jellyfish don't have this ability. For a more accurate image, you have to visualize having just one arm and one leg, but evenly spaced around the body. There is no way you can move such an arm or leg left or right—these directions simply don't exist in your world.

So it is with jellyfish: they don't know the difference between their back, belly, and sides. They have only one central axis, their axis of motion. It is as if jellyfish inhabited a one-dimensional world in which there were only forward and backward.

But hundreds of millions of years ago, one of our radially symmetric ancestors, probably resembling a modern polyp, fell on its side and started crawling on the seafloor.[15] In this moment, something curious happened—it sounds almost like a brain teaser. The polyp already had a front side—the side facing the direction of motion—and therefore a rear side too. But having fallen on the seafloor, it also realized that it now had a belly facing the ground and a back turned upward. Once you have both sets of axes, you get the third one—left and right—for free. The creature was no longer radially symmetric around a central axis—its sides were not the same anymore. Instead, it was symmetric across a plane that divided the body into two mirrored parts. This kind of symmetry, bilateral symmetry, was to become a trademark feature of all the animals to follow—they are collectively known as bilaterians, which include us. By falling onto the seafloor, the animal discovered a new dimension of motion: it could now move not just forward, but also differentiate between left and right, and so wiggle from side to side.

What could possibly be the reason for such strange behavior? It would have seemed as bizarre to an upright-standing polyp as the folding adventures of the sponge larva would have seemed to an adult sponge. Food comes from above, so why on Earth go sideways?

Food does indeed fall onto polyps from above. The thing is, this food, chiefly edible organic detritus consisting of dead microorganisms, had been snowing onto the seafloor for literally billions of years. By this point, it had compressed into a thick nutritive mat that padded the seafloor. It is thick even today: about 15 percent of all the world's bacteria are trapped in seafloor sediment.[16] Food, therefore, was everywhere, and you didn't necessarily have to sit and wait for it to fall on you—if you had a way to crawl over the surface of the mat, you could continuously scrape nutrients off the seafloor. A polyp, with its mouth pointed upward, was wholly unaware of this source of nourishment because it had no way of accessing it. But once it fell on its side, it became a different creature, one that was much more concerned with what was below than what was above—a *worm*. It was as if a one-dimensional creature transformed into a two-dimensional creature that gathered food from a surface, not just a straight line.

This, however, was not yet a typical worm of our time, like an earthworm. The main difference was that it had a closed gut—the same kind as a jellyfish or a polyp; whereas earthworms (as well as most other bilaterians, ourselves included) have what is called a complete gut, or through gut, meaning a gut with two openings on two sides of the body.

What's the big deal about the complete gut? For a polyp or jellyfish to feed, they need to put food into their mouth, close it, digest the food, absorb what can be absorbed, then open the mouth and release the undigested scraps. If, like a polyp, you sit and wait for food to fall on you, or, as a jellyfish, you swim and capture food in the water, it might not matter too much whether your gut has two openings or one.

But for worms, it is a game changer.

An earthworm, with its two gut openings, is built to dig, like a fish is built to swim. It feeds by constantly swallowing soil particles, absorbing any available nutrients, and expelling them on the rear side as it continues to burrow. This is not possible with only one opening—you can't continuously move and expel particles of soil in the same direction. So the first worms, those polyps that tumbled onto the seafloor and started wiggling around, could not yet dig as today's worms do.* That meant that although they turned to the seafloor as their source of nutrition, they could not fully access the vast reserves of organic material trapped in the nutritive mat, which at this stage was so dense it was almost impenetrable. The only thing they could do was scrape nutrients off the surface.

And then, at some point, the closed pouch of the worm's gut broke through on the other side, giving its body the canonical configuration of a bilaterian animal—a tube. The moment food could move through the body in one direction opposite that of its direction of motion, the worm acquired the ability to burrow into the nutritive mat. And burrow it did.

---

*Some modern worms, most notably *Acoelomorpha*, still look and function like that, and they might be our best bet at visualizing our ancestors at this stage of evolution.

The Rubicon

The most ancient known animal fossils belong to the so-called Ediacaran period (635 to 541 million years ago). They are rare and obscure, but they give us a glimpse into the earliest animal life on the planet. We can assume that during this period, there were swimming jellyfish and various larvae in the water, but they left no trace in the geological record. Most of the Ediacaran animals that did leave fossils had two things in common: they lived at the bottom of the sea, and they could not burrow.[17] Some members of Ediacaran fauna were sessile filter feeders like sponges, whereas others appear to have been grazing flat worms. We know virtually no animals from this time would live above or below the seafloor. As far as we can tell, animal life was mostly happening on a flat surface.

Then, 541 million years ago, something happened. During the Cambrian period, which followed the Ediacaran period, in just minutes on the evolutionary clock (some twenty-five million years), the ocean was filled with myriads of creatures of unprecedented complexity and unseen diversity. Suddenly, different animals started popping up everywhere—not just on the seafloor, but above, below, and soon even on dry land. All the major groups of animals recognizable to us today appeared during this extremely short phase of evolutionary history known as the Cambrian explosion. The contrast between the fossil records from the Ediacaran and Cambrian periods is astonishing, as if a sharp divide separated them in the timeline of life on Earth. On one side of the divide, paleontologists find rare bits of odd animals; on the other, piles of trilobites, tons of worms, and endless shells, segments, legs, teeth, studs, and antennae almost anywhere they look. The Cambrian explosion is a pivotal moment in the history of the animal kingdom and of planet Earth as a whole.

Nobody is quite sure what caused this. Some scientists actually deny the "explosion" altogether: the dramatic change seen in the fossil record, they say, could be explained simply by the appearance of hard shells and exoskeletons, as in trilobites, which are preserved immeasurably better than, say, the soft bodies of jellyfish.[18] Most experts do believe that the explosion was real, but disagree on the causes. One camp argues that it was triggered by a rise in oxygen levels, meaning an increase in energy availability. Another camp says that what lit the fuse was the invention of predation. Both of these ideas share the age-old logic of eukaryotic evolution: access to more energy begets an expansion in complexity.[19]

I prefer yet another version, which also revolves around food and complexity, although it may not be the most mainstream explanation. According to this idea, what triggered the Cambrian explosion was the "anal breakthrough"[20]—the transformation of a closed, single-opening gut into a complete gut with a mouth and an anus on opposite sides, which turned worms into tubes and gave them the ability to burrow into the sediment. These newly tubular organisms plunged themselves into this untapped trove of food with abandon, gorging on billions of years' worth of nutrients, like a mouse discovering an abandoned cheese factory. They multiplied prodigiously and rapidly diversified, evolving into scores of different worms, all harvesting nutrients from underneath the seafloor. Soon enough, this became the main thing animals did. And now, they could move not just in a straight line and not just on a surface, but in all three dimensions. The animal kingdom had finally regained the freedom of movement familiar to its single-celled ancestors but stifled by the early days of multicellularity.

These burrowing worms then untapped vast nutrient resources at the bottom of the sea and expanded the range of available habitats by enriching the ground below with oxygen and the water above with nutrients. New animals—mostly variations of the tubelike worms—quickly filled the newly available niches above and below the seafloor.[21] In the millions of years to come, the battle for dominance within the animal kingdom would unravel among their subgroups: from Teutonic, armor-clad arthropods to creepy five-armed echinoderms, from semiliquid, Dali-esque mollusks to springy, bouncy chordates. Chief among the latter would soon be a type of backboned worm called fish.

I like this anal breakthrough theory for several reasons. First of all, it's pretty funny.* But a more important reason is that it fits into the overall trend of animal evolution.

This trend is a reconquest of the world. A multicellular animal is a hypercell, an organism of a new scale, which must face all the challenges that had already been faced by single cells.

---

*An added twist is that the first phase of the Cambrian is called the Fortunian, so "anus of fortune" jokes abound.[22]

Consider the cell membrane. The reason an individual cell must be contained in this sealed capsule is to keep all its proteins, DNA, and RNA together in one place and to control the entry and exit of other substances from the environment. A multicellular body has something very similar called an epithelium: a flat, leakproof sheet of cells tightly bound to each other and glued onto a padding of collagen. Such epithelia line our gut cavities, blood vessels, and nasal passages, forming a sealed border through which only select chemicals can pass and preventing hormones and enzymes from leaking out. To an animal, it's essentially the same thing as the membrane is to a cell but made of many cells rather than many molecules. Epithelia are believed to have been invented by none other than sponge larvae—this must have been a precondition for any further animal evolution, just as the cell membrane was a precondition for much of the evolution of life on Earth.*

Consider also how similar the transitions between animal body plans—like the one from a sponge larva to a polyp to a worm—are to the transitions between cell types, like from archaea and bacteria to eukaryotes. In the single-celled world, you start with a simple bubble of a cell and end up with membranes within membranes within membranes, formed by folding, fusing, and splitting the original cell membrane into endless vesicles, vacuoles, and organelles. In the animal kingdom, you start with a simple bubble of a few hundred cells—the sponge larva—and by folding it step by step, like a sheet of origami paper, finally end up with cavities, tubes, tissues, and organs. Each of us retraces this evolutionary origami in our mother's womb, resembling, at various stages of our embryonic development, a sponge larva, a polyp, a worm, a fish.

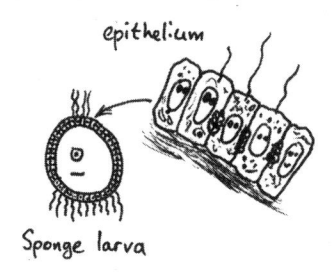

And so it seems to me that the reason for this Alexandrian moment in the history of the animal kingdom, the Cambrian explosion, must also have to do with the multicellular reconquest of the single-celled world. And at the very least it seems poetic that what would finally propel the animal kingdom to the pinnacle of natural greatness is the rediscovery of motion.

---

*Canonically, epithelia used to be considered one of the principal things that sponges did *not* have, as compared to all other animals. But on close inspection some species of sponges do show fully formed epithelia as larvae.[23]

Sponges don't move at all. Jellyfish can move in only one dimension. Ediacaran worms moved in two. Burrowing, enabled by the complete gut, opened up the final, third dimension of motion. Even if this wasn't *the* reason for the ascent of animals, it certainly set the rules of our world of moving things—a three-dimensional way of life to which our bodies and brains adhere to this day.

Origami rule number 1: never, ever, under any circumstances, cut or tear through the paper. But a true revolution is always a breakthrough.

Chapter 6.
Land!

If you cry, "Forward," you must make clear the direction in which to go.
ANTON CHEKHOV

Modern times. A predator lurks at the bottom of a pond, stepping carefully through the silt on its six jointed legs. It has a sleek, symmetrical, slender body encased in armor, spiky and sectioned into segments with geometrical precision. Two mother-of-pearl half-spheres form a shiny helmet on the predator's head—those are its eyes, or rather thousands of eyes looking simultaneously in all directions, each of them a microscopic hexagonal telescope. The predator looks as if it were molded from a titanium alloy smeared in mud for camouflage. Its movements are light, almost weightless. Suddenly every one of its many joints jolts, like the limbs of a string puppet. The lower lip, covered by a mask over the front of its head, swiftly extends into a sharp hook, revealing the predator's jaws, which in an instant clench in a death grip around the spine of an unsuspecting salamander.

The amphibian helplessly squirms in a last-ditch attempt to save itself. For a moment, it has a chance to wiggle out of the predator's clutches. If it does, it might grow up and emerge from the pond to see that fabled world—dry land—with its own eyes. But it does not. It is the six-legged predator that gets to see land instead. It takes a week, or maybe a month, but eventually the predator grabs on to a straw and climbs upward, toward the light. When it breaches the surface of the water and emerges into the merciless dry void of the atmosphere, it freezes, as if baked by sunlight onto the straw it just climbed.

At this moment it seems that the creature's path has come to an end. But suddenly the back of the predator's dried-out armor cracks, like an overripe pod. Slowly, over an hour, a new animal pulls itself out of the old shell. It comes out head first, pushing against the shell with its legs, then takes a break as if to rest, and finally, in a single push, contracts its entire body, breaking free of its underwater skin and unraveling four magnificent translucent wings. The day before, it was creeping through silt at the bottom of the pond; now, it takes off in flight, as comfortable in open air as it used to be in water. When the predator hovers over the pond, the front and the back wings beat in turn; when it bolts toward prey, wings synchronize for maximum acceleration. With its graceful, shining body, its polished, lightning-speed movements, and its all-seeing eyes, this creature shows a mastery of terrestrial life that a lowly salamander could have never hoped to match.

But then the predator's life comes to an abrupt ending. With its thousands of eyes, it does see all around, but it is not very good at noticing stationary objects. So it flies too close to something enormous and white and fails to react when, in an instant, the giant turns, tightens its muscles, and snaps its beak like a pair of sharp pincers. No fight, no warning—one of the most frightening and perfectly designed creatures of living nature is defeated in a single crunch. Tucking one leg under its feathers and absent-mindedly watching over the water, a heron towers over the pond. It just had a dragonfly for breakfast.

This story is fully realistic,[1] and we can use it as a metaphor for the entire relationship between the two great groups of terrestrial animals: insects and vertebrates.

A long time ago, in the Paleozoic era, it was insects who first found their way onto dry land. These almost perfect animals walked over the Earth's continents in a resolute march of conquest, adapting to every possible condition, occupying all available niches, both herbivorous and carnivorous, defeating any competition with their armored bodies, light wings, agile limbs, and infinite array of ingenious body designs. Only the arrival of vertebrates, with their unimaginably large jaws, claws, and beaks held up by a rigid spine, challenged the hegemony of these creatures on land.

At the scale of life on Earth, it hadn't been that long since both of these groups were one—both of them descend from the same tubular worms that blasted through the Cambrian explosion in the previous chapter. So if we started from the same place, and insects got there first, why didn't they push us back into the ocean when we attempted to invade their land?

"Every battle is won before it is ever fought," as famously put by Sun Tzu in the *Art of War*. The war between arthropods and vertebrates could have been the fiercest battle the animal kingdom had ever seen. But its outcome was decided long before the two sides reached the battlefield. Let us roll the clock back and survey the lay of the land as it was just prior to the arrival of animals.

And the Earth Was Waste and Void

Half a billion years ago, it made very little sense for any animal to crawl out of the ocean—there was nothing there except rocks. For land to become attractive to animals, it first had to be colonized by plants.

The first organisms to venture outside of the water were probably cyanobacteria peeking into open air at the water's edge.[2] It took eukaryotes much longer to get out of water, but when they finally did, with each step they changed the world around them, taking control of the air and the soil, rolling over the entire planet, and ultimately transforming dry land into a true eukaryotic empire, with its green jungles, fluttering wings, and silky furs. To build this Byzantium of nature on bare rock, standing on par with the ancient Rome of the ocean, was perhaps the greatest thing the eukaryotic domain has ever done.

The first eukaryotes to colonize land are believed to be neither animals nor plants but fungi—more precisely, lichens.[3] A lichen is a fungus filled with algae, basically a flat mushroom with a core of photosynthesizing cells. (*Algae* is a loose term used to describe many photosynthetic groups, and different kinds of them could be incorporated into the mushroom.) A lichen is a symbiotic creature—the algae produce the food, whereas the fungus protects them from dehydration, the greatest algal vulnerability outside of water. This arrangement possibly started with fungi invading a colony of cyanobacteria somewhere between water and land and taking control over their photosynthetic production while at the same time shielding them from water loss and so allowing further expansion onto land. The dry land environment back then was almost as harsh as outer space—not only was it by definition dry, but it also was unprotected from the sun's UV radiation by the ozone layer, as it is today, which would take a while longer to form. Lichens, however, are extraordinarily stress-resistant creatures. Today, they can live on sunburned rock without external sources of food or even water and have been shown to survive for weeks outside of an orbiting spacecraft.*[4] This resilience is what allowed them to pioneer life on land and prepare it for the arrival of others. Their most important contribution may have been simply the fact that they were there: by eroding and enriching the barren surface of dry land with their own organic material, these organisms set about creating what today we know as soil.

The great leap onto land taken by the next kingdom of eukaryotes—plants—occurred about 450 to 500 million years ago.[5] Before it could take place, however, plants had to solve several problems, each of them somehow connected with water—dry land is dry.

First, plants had to figure out how to gather water to begin with. Today, they achieve that using a specialized organ called a root. But the earliest land plants were rootless, creeping sheets of green cells—the best reference today might be a liverwort. Instead of roots, they employed symbiotic fungi, which attached themselves to their underside, grew deep into the soil, and pumped water upward in exchange for a share of photosynthetic food.[6] Only later in evolution did plants supplement these "fungal roots," or mycorrhiza, with their own actual roots, and the two still usually coexist in modern plants. So not one, but two fungal symbioses played a key role in the establishment of land life: first lichens, which prepared the soil for the arrival of plants, then mycorrhiza, which almost literally dragged them out of the water.

The next problem was water loss. Plants, unlike animals, cannot simply move toward a source of water or take refuge in the shade when they are dehydrated—they have to stay in the same place, motionless, no matter what happens. For this reason, land plants are typically coated with a waterproof waxy layer called a cuticle,[7] which they supplement with specialized "mouth cells," or stomata, which open and close depending on temperature and humidity and can seal a plant shut or unseal it to allow gas exchange. The analogy with the mouth is pretty accurate, since stomata let in carbon dioxide, the raw material for making food during photosynthesis, serving the same function for plants as a mouth does for animals.

---

*Sadly, their tendency to absorb any trace of moisture makes lichens extremely sensitive to air pollution, and the one thing they cannot tolerate is the concrete jungle of big cities.

Dehydration of leaves is one thing—but a much bigger problem is the dehydration of spores, tiny particles that ancient plants used for dispersal. There was no problem if spores were dispersed by water. But to truly establish themselves on firm ground, plants needed a way to spread by air—by wind. The issue is that whatever it is that a plant may be releasing into the wind—spores, or later in evolution, pollen—has to be a very small piece of living tissue, which means, basically, a miniscule droplet of water. For such a microscopic droplet, being surrounded by air on all sides normally means instant evaporation and death. Preventing this seems to border on the impossible—and yet plants figured out a way. The solution was a biomolecule named sporopollenin—a chemical of extraordinary strength and stability, in a way more similar to human-made plastics than to the typical components of a cell.[8] Sporopollenin forms a tiny, hermetically sealed waterproof casing around spores and pollen, a spacesuit that allows them to travel by air. The cells inside of the granule hold the only key to unlocking it: specialized enzymes that dissolve sporopollenin upon reaching their destination.* Out of all the solutions that plants found to their problems, sporopollenin may have been *the* evolutionary invention that unlocked land life.

All this relentless problem-solving by the founding fathers of our land Byzantium suggests one thing: for whatever reason, plants really wanted to get out of the water. (And fungi really wanted to help.) What could possibly be the reason to transition to such an unnatural form of existence, alien to all previous forms of life on Earth?

One could argue that the expansion of plants onto land was predictable simply because the ocean would sooner or later run out of space. But in fact, there is plenty of space in the ocean. That is not what limits the proliferation of underwater plants and algae. Plants are far more constrained by space on land than they are underwater.

In part, competition by plants for space on land is tighter because shadows are a bigger problem (underwater, light is more diffuse). This, by the way, is why land plants possess a fascinating device: the stem.[9] Underwater, stems

---

*Sporopollenin coats spores and, in flowering plants, pollen. A spore is a single-celled progenitor of a new plant, and it does not require a sexual partner—it simply dissolves its casing after landing in a suitable area and starts growing. A pollen grain is a complete male organism, usually three-celled, which seeks a partner, the female organism that inhabits the pistil of a flower. Sporopollenin plays the same role in both cases.

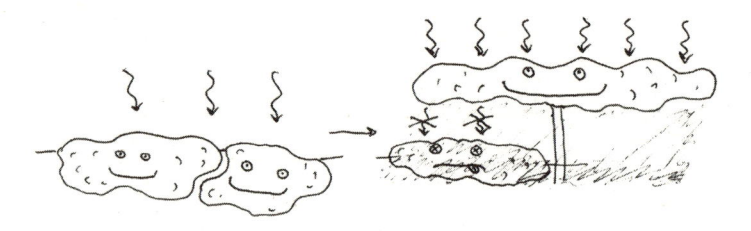

You only need a stem if your neighbor has a stem.

(as well as roots) are practically useless. But for a plant to succeed on land, a stem is often critical. Why? Because if everyone else has one, and you don't, your neighbor gets all the light, and you get shade. It's not that a stem, by itself, helps plants photosynthesize or brings them meaningfully closer to the sun. It simply lets one plant get ahead of another. Vertical growth among plants—the sort of growth that created the Amazon jungle and the Siberian taiga—is an arms race. The bigger the stem, the higher the chances of maximizing sunlight and casting a shadow over all your competitors. Sometimes, as in the case of a tree trunk, this inherently useless structure requires colossal investments of biomass and takes decades to build up. From the vantage point of the plant kingdom, it is a risky and expensive tactic whose long-term success can be justified only by cutthroat competition. How competitive must space on land be to justify investment in a redwood tree one hundred meters in height?

No, the reason land turned green is not that plants had nowhere else to go. Terrestrial habitats must have offered plants something else that attracted them.

At the most basic level, a photosynthesizing organism needs three things: water, light, and carbon dioxide. It needs other things, too, such as nitrogen and phosphorus, but water, light, and carbon are the big ones. As we've seen, moving onto land creates major problems with water. But in terms of access to light and carbon dioxide, land life is far superior. With less diffusion of light in the air means that a given leaf can harvest more sunlight than it can in the water. More carbon dioxide (which is not very soluble in water) means a glut of raw material. These are the two factors that lured the green kingdom onto land: light and carbon. To ancient plants contemplating this transition, the harsh and unexplored terrestrial habitats offered a lavishly resource-rich atmosphere. This was the reason plants mastered life outside of water—and we animals followed suit.

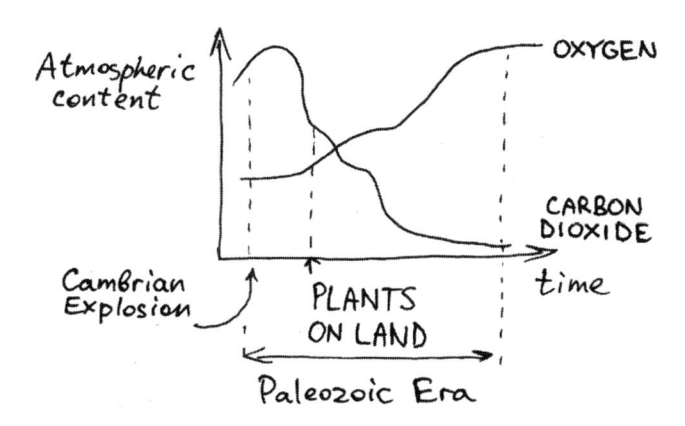

[Figure: Hand-drawn graph showing Atmospheric content over time during the Paleozoic Era. Oxygen curve rises while Carbon Dioxide curve falls, with markers for "Cambrian Explosion" and "Plants on Land."]

What plants wanted was access to raw biomass: the sheer gigatons of carbon floating in the dry air above the ocean.

And so, eventually, plants solved all their water-related problems and swarmed the continents like a green avalanche. Early in the Cambrian period, back when land was still barren rock, oxygen made up about 15 percent of the atmosphere and carbon dioxide about 0.6 percent. By the end of the Paleozoic era, when land was covered with dense jungles and mighty forests, oxygen levels had more than doubled, while carbon dioxide levels had dropped *seventeenfold*, to a mere 0.035 percent.[10]

Basking in sunlight, plants gobbled up almost all the carbon from the atmosphere, physically bringing it down to Earth, precipitating it into their bodies. As a side effect, they pumped the air with unprecedented quantities of oxygen, which peaked at a whopping 35 percent (today it is about 21 percent). As we shall see, this will be the last time the planet sees such a glut of biomass on land and oxygen in the atmosphere.

In short, plants came onto land for the light and for the air.

Animals came onto land for the plants.

A Horse-Sized Ant

If I had a time machine, the first ancient creatures I'd visit would not be dinosaurs, but giant dragonflies of the Paleozoic. Imagine a dragonfly in all its grace and creepiness, but ten times larger, about the size of a seagull. Picture a world in which this animal is the king of beasts, the supreme predator. The air is warm and tropical. Land is covered with dense jungles of giant ferns and horsetails (it would be hundreds of millions of years before flowering plants appear).[11]

This describes the Carboniferous period of the Paleozoic era—you could argue that this is when life on land peaked in its copious prosperity. Landscapes were lush and teeming with activity. Propelled by extraordinary levels of oxygen, insects dominated the environment, feasting on plants, fungi, and each other.

But already at this stage, new contenders for land dominance appeared on the scene. They were vertebrates, our ancestors. Soon, they would reach truly cyclopean sizes, spread into every corner of the habitable world, and ascend to the top of the food chain wherever they set foot. The dragonfly was never to be an apex predator again.

If back in the Paleozoic era insects were such advanced creatures, how could they have failed to defend their territory?

On the surface, the reason for this is straightforward, although I think it worth pausing to contemplate it for a second. We vertebrates are the *largest animals in the world*. We tend to think of everyone else as small, but really, there has never been any other group even remotely comparable to us in the amount of organic matter that we drag around on our bones. By the standards of the animal kingdom as a whole, we are all giants—even rats.

But why us, and not anyone else? Vertebrates are just one of the many branches of the animal kingdom, and invertebrates comprise all the rest, and yet examples of invertebrates preying on vertebrates are exceedingly rare simply because we are usually the largest ones in any ecosystem. How come no one else was able to wrestle dominance away from us at least some of the times? Dragonfly larvae in freshwater ponds, which I described at the beginning of this chapter, are a notable exception: they do sometimes tackle a small frog, fish, or salamander, but nothing larger than that.

The key part of the reason is our vertebrate skeleton. It is especially important if we are talking about life on land. In the water, relatively large animals can get by without a rigid frame for their body—for example, some octopuses are big and powerful enough to tackle even a sizeable shark. But without water to help counteract gravity, anything that wants to lift itself off the ground even slightly requires a scaffold; just like in architecture, any sufficiently tall building requires reinforced concrete so it does not collapse under its own weight. That same octopus, fearsome as it is in the water, becomes a helpless pile of slime when removed from it.

Insects, however, *do* have skeletons, as do all arthropods—a broader group that also includes crustaceans, spiders, and millipedes. It is part of the reason arthropods are, without a doubt, the most successful group of animals on the planet, both in biomass and in variety.[12] A rigid exoskeleton, made mobile by supplementing it with jointed legs, has been their trading card since the Cambrian explosion—think of a seafloor-roving trilobite, probably the most famous ancient invertebrate. An exoskeleton offers a powerful advantage beyond supporting body weight: it simultaneously acts as armor. It is almost impossible for an animal of comparable weight to bite through an arthropod's exoskeleton. If you watch two insects fight, they rarely attempt to do that. Instead, they try to bite between the joints, the only exposed soft part of the body. The exoskeleton also protects arthropods from other forms of distress, such as dehydration, which becomes important once they venture outside of water. Essentially, this armor around their bodies is the entire bet of the arthropod lineage, and over the past five hundred million years it proved to be an extremely successful bet.

So why, then, did arthropods fail to stand up to our vertebrate ancestors slowly creeping out of the water into their already established land empire? It might seem like science fiction today, but back then it wouldn't have been that far-fetched to envision an alternative turn of events. Imagine an insect the size of a horse encountering a fish struggling to make its way through mud or a shallow pool—probably the most realistic version of how vertebrates got their first taste of dry land.[13] If insects—ants, for example—were the size of horses, there's little doubt that there would be no room for vertebrates on planet Earth. Our outer soft tissues are not protected from attack by these killing machines in any way, because our internal skeleton—endoskeleton—does not shield us.*
But it does give us a different powerful advantage: we don't need to molt.

The problem with an exoskeleton is that it stands in the way of growth. It must be shed every time an insect wants to enlarge in size. To do that, it must remove the hard shell, grow as quickly as possible, and then solidify a new shell around itself. So insects, like all arthropods, grow in bursts and are extremely vulnerable during these short periods. This puts a cap on how large an insect can be—no matter how strong its exoskeleton, it has to be removed for growth. To imagine an insect the size of a horse, we must first imagine an insect the size of a pony and then picture what would happen if that pony discarded all of

---

*The skull is one exception, acting almost like an exoskeleton for the head.

its bones: just like an octopus on dry land, it would turn into a helpless pile, a ready-made meal for all the predators in the neighborhood. So an insect needs to remain small enough to survive *without* a skeleton for a period of time. Even for a rat-sized insect that is difficult.

Our skeletons, on the other hand, grow together with our bodies. They do not limit our growth in the same way that external shells do, and when they need to be restructured, they are continuously dissolved and rebuilt by living cells inside of them. So our bodies can be enlarged indefinitely, far beyond what is possible with a periodically molting exoskeleton of an arthropod. It is this difference in the positioning of skeletons—a choice once made by two similar wormlike creatures in the Cambrian ocean—that determined which one of us would end up under whose feet.

But we were not always *this* large, and they were not always *this* small. Maybe arthropods could never be the size of a horse—at least not on land—due to the basic configuration of their body. But we know that they could be the size of a seagull, as dragonflies once were in the Carboniferous. A swarm of such creatures, sufficiently aggressive, would have surely posed a serious threat to our fish-sized ancestors attempting to make inroads into their territory. What happened to all those giant dragonflies, so that today, we no longer consider any insect a serious threat?*

This particular question happens to be one of my favorite questions in biology—it is why this chapter is all about dragonflies. It takes me back to my undergraduate years at St. Petersburg State University.

Zenith of the Insects

In Russian universities, exams are usually oral. Your entire course grade is decided by a single interview during which you could be asked anything. Students are admitted to the classroom staggered in groups of four or five people. You draw a "ticket" (like a lottery ticket), a narrow sheet of paper laid facedown on the professor's desk. There are usually ten or twenty of them. Each ticket lists two or three topics from the course. You are given some time to prepare, as more students enter to draw their tickets. When ready, you raise

---

*Statistically speaking, we probably should: the animal that is most dangerous to modern humans is unquestionably the mosquito due to its unfortunate role as a carrier of many diseases.

your hand to get interviewed. For big courses, a few professors would interview the students, so you'd often have someone you had never met decide the entire outcome of your semester. They listen to what you have to say about the questions on the ticket but usually wave the prepared questions off, asking about whatever they find interesting to figure out how much you know and how well you understand it. There's an element of lottery and a gamble and plenty of room for bias and subjective judgment, which is why we don't do this in US universities. But nothing else in my life taught me as much about speaking clearly, thinking on my feet, answering questions, and arguing my point as those oral exams did.

One of the most difficult courses (and exams) I had to take during my freshman year was invertebrate zoology, widely considered a hardcore rite of passage—something like anatomy in US medical school. The way I passed this exam involved a good deal of prep but also a good deal of hustle, which is why I am especially proud of it.

By then I was an expert at oral exams. I planned my strategy like a spy operation with a hint of voodoo—for example, for my entire college career, I have only used my left hand to draw the ticket. (Exam-related superstitions are a whole genre of Russian urban lore—some people go as far as putting their grade books in the freezer when grades are good, whereas others open their windows at midnight preceding the exam and call out for *khalyava*, a deeply Russian term for success by pure luck without effort.) Because there were a few professors, each becoming available whenever they were done interviewing the previous student, and because within a certain limit, you could volunteer to go when ready, you could game your timing to get to the "right" interviewer. I knew exactly which professor to aim for. This wasn't because we had a special relationship—in fact, I had never met him. The reason was that this particular professor's pet topic—Carboniferous dragonflies—had been reported to me by a group of students who had taken the exam the previous day.

It took a bit of skill to dodge all the other professors to get the one I needed when he became available. Once we sat down, I started talking. I knew that the questions on the ticket were irrelevant—my goal was to slowly shift the conversation toward the dragonflies.

The task was made easy by nemertine worms, which featured in one of my official questions on the ticket—those worms can reach more than a meter in length and present a wonderful opportunity to make an offhand remark: "Isn't it sad there aren't many giant invertebrates around, such as, say, *really big insects?*"

Now the professor had the twinkle in his eye that I was waiting for. He struck right where I was hoping he would. "What about," he said, "the giant dragonflies that were around during the Carboniferous? What was different then?"

I took a moment to fake confusion. I looked at the ceiling, I rubbed my forehead and apologetically admitted that I didn't know for sure. But perhaps, I said, it had something to do with their tracheal breathing. Maybe it set the limit on how big insects could be, so I could speculate that in the Carboniferous, when there was more oxygen in the atmosphere, the limit might have been higher?

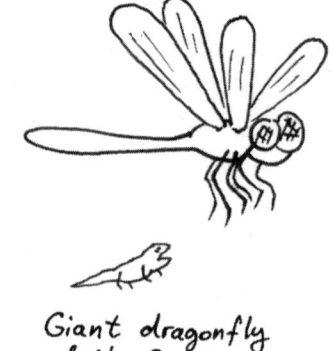

I passed the exam with flying colors.

Whereas we vertebrates breathe using lungs or gills, insects breathe using a network of air tubes permeating their bodies—tracheae.

These tubes act as capillaries for air, delivering oxygen straight to the organs and branching extensively around muscles. The "veins" on insect wings are also built around tracheae. All these air tubes open directly into the atmosphere through valves reminiscent of plant stomata. As a whole, insect breathing actually resembles plant breathing. Neither insects nor plants pump air by inhaling and exhaling, but rather by allowing gases to passively diffuse inward and sealing themselves shut if it gets too hot or too dry.

On one hand, this is a great idea. Dry tracheae minimize water loss—this might have been a major reason why this system of breathing was first adopted. Additionally, oxygen is poorly soluble in water—so supplying organs directly with air seems like a much more efficient idea than trying to squeeze oxygen into a liquid medium, like blood.

But tracheae only work for small bodies. Not having active ventilation means that insects have a size limit beyond which oxygen takes too long to penetrate deep into the tissues in sufficient quantities. This passive breathing keeps insects from becoming too large, and it makes them highly dependent on oxygen levels in the atmosphere.

This is the reason the bird-sized dragonflies of the Carboniferous were possible then but not now: there was more oxygen in the atmosphere, which meant that the upper limit on insect size was higher than it is today.[14] After the Carboniferous, oxygen levels started to decrease, and insects never returned to their former glory. They had made a bet on passive breathing—like plants—and lost.

Vertebrate breathing, by contrast, is active, using a lot of energy at every stage. Oxygen supply is ensured either by ventilating lungs or, as in fish, by moving gills through oxygen-rich water. Oxygen is then distributed by blood, which is actively pumped by a heart. Although oxygen moves slowly through a liquid, the liquid itself moves fast. Although oxygen is poorly soluble in water, we manage to pack thirty to forty times more of it into our blood thanks to hemoglobin, a protein that sucks oxygen into a sort of chemical force field as it moves through the body.[15]

At smaller body sizes, the two breathing strategies—the passive and the active one—may be comparable, but our kind is much more scalable than the insect kind: it allows us to maintain organs far removed from the lungs and therefore opens the door for much larger body sizes. Even as oxygen levels dropped in the atmosphere after the Carboniferous, we were able to continue growing and growing and growing.

The era of giants, the Mesozoic, is visible on the horizon.

The Breath of the Paleozoic

You can tell that insects were designed in the Paleozoic era.

Insects spent their formative years watching land become greener and greener and the air richer and richer in oxygen content. Their main existential

Insects and worms arrive on land

problem was water. So instead of choosing the potentially available option of pumping air back and forth into their bodies, they bet on the system of breathing that best preserved moisture and used the least energy.

During those first few millions of years of carefree life, insects had no reason to be large because there wasn't anything larger than them. Under such circumstances, their exoskeleton provided effective protection against any attack, and their fine mesh of tracheal fibers easily infused their bodies with oxygen. What's large and what's small is really a matter of point of view. Insects were much larger than, say, roundworms, a mostly microscopic group of animals that might have first made it onto land as parasites of arthropods.[16] To animals, large bodies are what large stems are to plants: you need one only if your neighbor has one. Before vertebrates were around early in the Paleozoic, insects couldn't have imagined in their wildest dreams what sizes are possible if you have a spine. No reasonable animal of the time could have seriously considered the possibility of a dinosaur, or a blue whale, or even a cat. An animal body, after all, is just a supplementary device meant to facilitate the transition of its germ line into the next generation—you don't need a lot of body weight for that. We humans don't consider ourselves particularly large, but seen from this historical juncture, our bodies—*kilograms* of living matter—are a laughable proposition. Only when vertebrates started actually towering over insects did the advantages of this possibility—and the inescapable drawbacks of the exoskeleton and the tracheae—became apparent. Insects got stuck in the Paleozoic. As stated by Sun Tzu, the outcome of the battle was determined long before it was ever fought.

In fact, the battle never even had to happen. By separating themselves into utterly distinct size ranges—large and small—vertebrates and insects found what may have been the only possible way for them to coexist.

On one hand, you could say that we vertebrates defeated insects in the greatest cold war of all time. We entered their house and, without ever facing a head-on collision with similarly sized insects, firmly established ourselves at the top of the ecological pyramid. (This is one graphical representation of a food chain: at the base of the pyramid sit plants, above them herbivores, then carnivores, and above all of them stands a man with a big harpoon—at least in old textbooks that had no problem with this imagery.)

But on the other hand, I'm not sure insects would agree that they ever lost anything in any war, hot or cold. Today's insects as a whole are not terribly interested in vertebrates and seem satisfied with the size range they ended up with. Why would they try to challenge us? Vertebrates remain a minority in the world of beetles, flies, and caterpillars. Insects are perfectly content with their lives and are extraordinarily successful. To be sure, many of them are eaten by vertebrates, but in most cases this is not their primary concern.

In fact, why should standing atop the food chain be considered a victory? It is actually much harder to be a carnivore than a herbivore. The further away from the primary source of energy—photosynthesis—the more energy is lost. The pyramid is only a symbol, which could just as well be flipped, in which case we would end up not on top of a pyramid but at the bottom of a pit. If anything, that would represent reality more accurately: the farther down, the farther away from the sun.

Man with a harpoon at the bottom of an ecological pit

The real story, insects would say, is their invincibility. To this day, nothing weighing twenty milligrams has the same agility, strength, speed, range, and raw, terrifying Darwinian ruthlessness as a wasp. Insects might argue that this, and no other consideration, was the reason we vertebrates were forced to invest in our ridiculous, bloated, absurdly expensive bodies—all to avoid a battle with them, the truly dominant group of land animals. How much time and resources does it take to raise just one fully functional new human? In that time, thousands of generations of insects are born, live their lives, leave offspring, and die.

Chapter 7.
When the World Ends

Not a prince to awake, but a dinosaur.
JOSEPH BRODSKY

Picture yourself as a dinosaur—any dinosaur—in its natural habitat. What other creatures do you see around you? I can almost guarantee that whatever you are imagining, you are either alone or hanging out with other big reptiles—maybe you are a T. rex taking a bite out of a triceratops or a stegosaurus looking at a pterodactyl in the sky.

Now back to you being a human, and let me ask you a question—where do you think *we* are in this picture you just imagined?

Of course humans as a species are a long way off. But our *lineage*—the chain of generations eventually leading up to humans—must have existed the entire time dinosaurs ruled the world, or roughly the extent of the Mesozoic era. Since we don't descend from dinosaurs themselves, there must have been someone else who had existed *alongside* them and later given rise to us. This dinosaur world you imagined? Our ancestors must have been there too. Yet somehow there's no mention of them in *Jurassic Park* or any cartoons of them on grade school stationery.

As we are about to see, during the Mesozoic era, when big reptiles ruled land, water, and sky, our ancestors were not only around, but living through arguably the most consequential period in their history. There were very few of them—so few that they almost went extinct, hanging by a thread for more than a hundred million years. The reason we don't have an image of them mingling with dinosaurs is because they retreated from dinosaurs by adopting a nocturnal lifestyle and disappearing from their view. They were small, stealthy, almost invisible creatures lurking in the shadows of daytime monsters, hiding from them in the darkness of the Mesozoic night.

But it wasn't always like this. In fact, these small, seemingly inconsequential creatures descended from some of the biggest, fiercest, most powerful animals that had existed before dinosaurs were around. And they wouldn't always be like this, either. Eventually, they not only outlive dinosaurs—they *replace* them.

In the previous chapter, we considered a head-to-head collision that never happened—the two dominant groups of terrestrial animals, insects and vertebrates, ended up coexisting in distinct size ranges without ever facing a clash on even terms. The greatest head-to-head collision that nature did witness—the battle of synapsids and sauropsids—occurred instead *within* our vertebrate clan. One might even say that it continues to this day.

Long before the great rivalry of synapsids and sauropsids unfolded, a distant aquatic ancestor of both of these vertebrate clans—a fish with especially muscular fins—had to crawl from water onto land. (This must be the second most iconic image of evolution, after the transition of apes to humans.) Fish were probably crawling from pond to pond or from lagoon to lagoon and gradually started spending more time out of water, after discovering that land was apparently swarming with fully snackable spiders, millipedes, and early insects.[1]

Gradually, these land-crawling fish modified their bodies to suit the new environment. Their fins turned into jointed limbs (a shameless rip-off, arthropods would surely say), and swim bladders (flotation bubbles in most fish) developed into specialized organs for breathing outside of water—lungs. Soon enough, amphibians—the first major group of four-legged, terrestrial vertebrates—proudly walked the Paleozoic land.

But despite their four legs, amphibians held on to a semiaquatic lifestyle. They still do—modern frogs and salamanders cannot reproduce away from a body of water, as if still standing one foot in the ocean from which their ancestors emerged. The main reason is their eggs. Amphibian eggs are not much different from the eggs of fish—both have to be placed in a wet environment, such as a pond. When frogs hatch from their eggs, they don't become adults right away, but rather tadpoles—also very fishlike. Only after these tadpoles bulk up can they then grow legs and crawl out of the water.

This was not a problem for amphibians during their Paleozoic heyday, which roughly coincided with the giant dragonflies from the last chapter. At that time, there was no shortage of swamps and puddles where amphibians could incubate their eggs, because the entire landmass around the equator was covered with a continuous belt of warm, humid rainforests. Amphibians established themselves comfortably throughout the world and often reached large sizes—*Pholiderpeton*, for example, looked like a mix between a salamander and a giant eel and with its long swimming tail reached up to three meters in length; *Anthracosaurus* resembled a large, somewhat froglike crocodile. Today's newts and toads are a mere shadow of this former glory.

But as climate became drier toward the end of the Paleozoic era, rainforests collapsed, and amphibians were thrown on the defensive. The boundless jungle splintered into isolated patches of forest surrounded by inhospitable terrain with no standing water.[2] This put severe pressure on amphibians, which were left with fewer and fewer places to reproduce.

In this increasingly dry environment, a new group of vertebrates gained advantage: amniotes. What gave them the edge over the amphibians was a specialized egg, which contained an internal membrane, called amnion, that sealed the developing embryo in a water-filled capsule within the egg. It's like carrying a fish from a pet store in a plastic bag. Since the water was sealed inside the egg, the egg itself could be kept in a dry place. This was the secret to amniotes' success. (Plants would no doubt insist that amniotes stole the idea

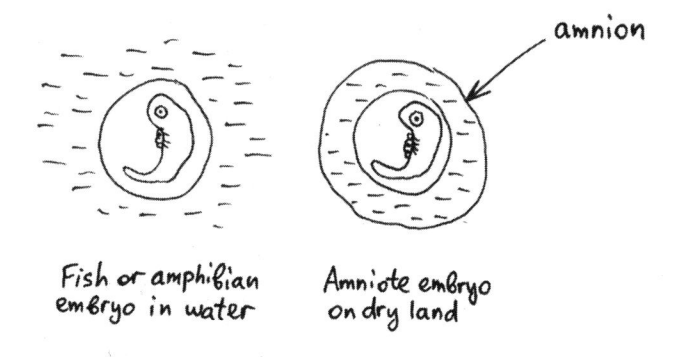

Fish or amphibian embryo in water

Amniote embryo on dry land

from them. The essence of an amnion is eerily reminiscent of sporopollenin, the key adaptation that enabled green plants to get out of water: recall that plants used this plastic-like chemical to seal moisture inside their spores, which enabled them to spread by air and colonize dry land.)

And so, as rainforests continued to shrink, amniotes began to displace amphibians, which remained tethered to increasingly rare bodies of water.

It is around this time—very shortly into the tenure of amniotes—that the great rivalry of their two clans begins.

There was not just a single line of amniotes. From early on, there were two separate groups of them, called synapsids and sauropsids. They had a common grandparent, a small, early amniote probably resembling something between a salamander and a lizard. So the first synapsids and the first sauropsids probably looked something like that as well—not much to see compared to the great amphibians that were still around and not much different from each other.

But there was something that gave synapsids an advantage over sauropsids. I would have loved to see this moment in history with my own eyes: to witness two similar creatures going head-to-head in pursuit of prey—maybe a juicy centipede—one of them emerging victorious. The advantage needn't have been a lot. A tiny initial difference in speed, agility, or strength could have been enough to start a snowball effect: a steadier diet of centipedes leading to larger numbers and greater diversity, small evolutionary advantages turning into other, bigger evolutionary advantages.

The most obvious explanation for the dominance of synapsids over sauropsids at these earliest stages of their rivalry lies in their skulls—it is the original, most ancient distinction between the two groups that we can identify.

Up to this point, vertebrates had heavy armored skulls that fully encased their heads. Synapsids lost some of the bone: they developed an opening in the skull behind the eye socket. The opening made the skull lighter and so must have made synapsids faster. Even more importantly, jaw muscles, rather than just connecting the lower jaw to the upper jaw, could now be threaded through the opening and attached further back, to the outside of the skull, giving them more power.[3] Another jaw-related feature that from very early on distinguished synapsids from sauropsids was differentiated teeth: incisors, canines, and, later, molars—each type of tooth useful for a particular function.[4]

These skull modifications not only made synapsids more successful at capturing prey—they were also great adaptations for eating plants. In fact, synapsids were among the first vertebrate herbivores.[5]

It's worth dwelling on this for a moment. Why weren't all the vertebrates eating plants to begin with? You would think that herbivory should be the most obvious strategy of all: plants are the most widely available source of organic matter on the planet. But in fact, most of this matter is extremely difficult to consume—try eating a log.

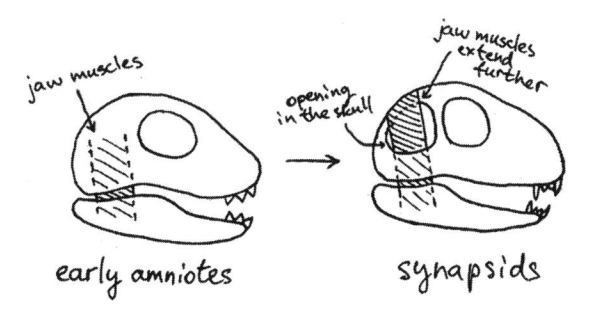

The reason is that animals, as a whole, lack enzymes to digest cellulose, the main component of a plant cell wall. If you squeeze water out of a plant, what remains is mostly cellulose, and this is what animals can't digest. What herbivores do instead is carry in their guts microorganisms that do break down the cellulose. Microbes ferment the plant, and the animal then feeds on the microbial slurry—this is true for herbivorous vertebrates and invertebrates alike. Such indirect feeding, mediated by microbes, is a complicated and energetically inefficient strategy, which explains why it wasn't the first thing vertebrates started doing as they moved onto land—it was much easier to eat bugs and, later, other vertebrates.

It is, however, the strangest thing that herbivory is such a hassle. In principle, a cow could digest cellulose without the mediation of any microorganisms as easily as we digest starch (chemically, these molecules are very similar). It just lacks an enzyme—a basic, ordinary enzyme possessed by many bacteria. Here, it seems, is an example of a perfectly solvable evolutionary problem with a great potential benefit. Why can't herbivores just evolve a cellulose-digesting enzyme of their own?*[6]

Although I can't say for sure, I think the reason is that by the time in their evolutionary history when animals started eating plants, they were too large and complicated to simply evolve a new enzyme. Evolving a new enzyme is a tedious, unlikely evolutionary task, which requires a brute force of natural selection sifting through billions of rejected versions along the way. It is something microorganisms and fungi are good at because they are simple and reproduce fast. But for an animal, which spends days or even months developing into a fully grown creature, it is practically impossible. The first animals—sponges—fed not on plants, but on bacteria, and did not need to break down cellulose. (Accordingly, we do have enzymes for breaking down *bacterial* cell walls.) Plants as a food source came into the picture a lot later, by which point it was much easier for animals to form an alliance with plant-digesting microbes than to evolve their own enzymes to that end.

But back to Paleozoic synapsids. In summary, it was a range of factors that established them as the number one clan of amniotes of their time—stronger jaws, better teeth, a greater range of diets, including a vegetarian option, and as

---

*Although it is textbook dogma that animals don't have cellulose-digesting enzymes (cellulases), there are in fact reports of a few cellulases produced by some nematodes and termites, rather than by their symbiotic bacteria.

a result of all this, greater size and metabolic rate. From their humble lizardlike beginnings, synapsids grew to become almost as large as the future dinosaurs—you could say they did the dinosaur thing before dinosaurs were around. Creatures such as the *Dimetrodon* and the *Cotylorhynchus*, for example, reached the size of an alligator and an elephant seal, respectively. The mega-herbivore *Estemmenosuchus* looked like a hippopotamus with a massive tail and Shrek-like horns and was undoubtedly harassed by mega-predators such as gorgonopsians, which resembled large saber-toothed cats with a slightly snakelike shape to their head.

Throughout these boom days of synapsids, there had been no indication of any ongoing rivalry—synapsids probably hadn't thought about sauropsids since they last squabbled over a centipede, shortly after the two clans had been established, and probably would have been surprised to find out the sauropsids hadn't gone extinct yet. By the end of the Paleozoic era, synapsids had fully pushed sauropsids into an evolutionary corner, where they remained—small, obscure, unremarkable, seemingly bound for the footnotes of future textbooks.

But the prime time of synapsids was rapidly coming to an end, as was the entire Paleozoic era, which began with the Cambrian explosion hundreds of millions of years prior. The next era, the Mesozoic, was to be the era of sauropsid revenge.

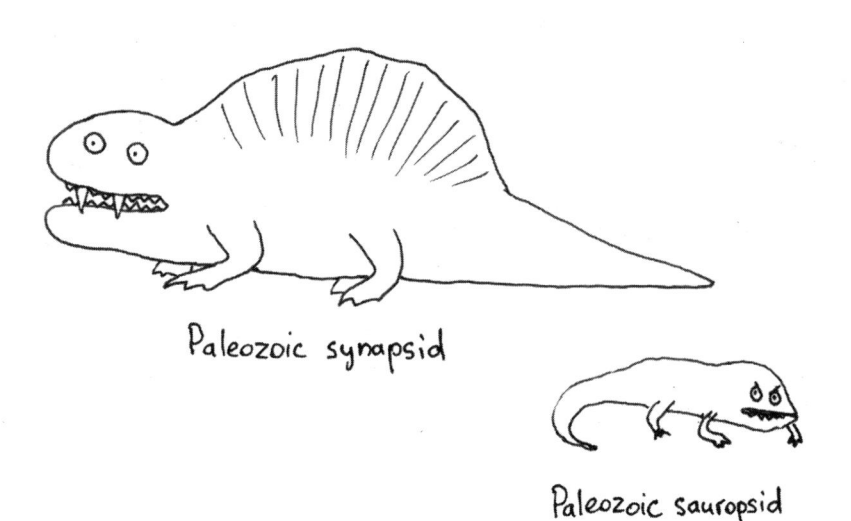

Paleozoic synapsid

Paleozoic sauropsid

Pompeii Times a Million

You can still see it from space—the snow-covered streaks of basalt, the hardened magma spilled over a territory about the size of Great Britain. The Putorana plateau in northern Siberia is a tombstone of the Paleozoic era—what remains of the apocalyptic eruption of a massive "province" of volcanoes lasting *a million years*. This event caused the extinction of a large fraction of life on Earth, known as the end-Permian extinction (named for the last period of the Paleozoic era, the Permian period), or simply the Great Dying.

It is hard to imagine such an event. My internal standard for volcanoes is set by *The Last Day of Pompeii*, a large historical painting by Karl Bryullov and one of the main attractions in the Russian Museum in St. Petersburg, where my parents used to take me as a child. The painting is epic in every imaginable way. It depicts the apocalypse: crowds aghast, statues toppling down the steps of a palace, and cherubic children and riders on white horses succumbing to the red-and-black inferno of erupting Mount Vesuvius. Very scary. Pompeii was destroyed in AD 79, so I look at the painting and imagine that this eruption of Vesuvius is still ongoing two millennia later; it will go on like that another *five hundred times* over; and it's not just in Pompeii but all over the European continent. That's the scale of disaster that the Siberian volcanoes unleashed at the end of the Permian period. Siberia is hardcore.

There is still a lot of debate about how exactly the massive eruption connects to the great extinction. There were species directly affected by the volcanoes, which perished quickly. The rest of life on the planet suffered indirectly. Some experts blame global warming caused by the release of underground gases. Others blame blackened skies and endless fires that apparently consumed the entire land mass—this can be seen in the geological record.[7] Marine species might have died out because of ocean acidification.[8] Yet another factor may have been methane abruptly released into the atmosphere, either due to physical reasons[9] or because of the proliferation of some air-polluting microbe.[10]

One way or another, the Paleozoic era was now over. Its megafauna had been consumed in the fires of the Great Dying. And among the main victims of this catastrophe were synapsids, who up to this point dominated land life so impressively. It was these enormous, energetically expensive creatures that ended up being the most vulnerable to any disturbance in the fragile natural equilibrium that kept them alive. They were the first to run low on food, water, and oxygen. And so, as the Paleozoic gave way to the Mesozoic, these specialist giants were wiped out, leaving behind only the species that were smaller but more resilient, unassuming, omnivorous.

Among these few surviving synapsids were listrosaurs, which looked something like extraterrestrial pigs.[11] They became surprisingly abundant in the first stages of the post-apocalypse recovery. Like mice in an abandoned house, these blunt-snouted creatures scuttled across the scorched earth, gulping the thinned-out air into their enlarged lungs. They foraged for food by digging with their unusually shaped skull (*Listrosaurus* means "shovel lizard") and might have even lived underground, like moles.[12] These creatures had existed before but shrunk in size after the Great Dying, from the size of a boar to about an average dog.

There was another, less noticeable group of synapsids that also survived the Great Dying: cynodonts, which resembled slightly serpentine badgers. Although no one could have foretold that at the time, it is these animals that were destined to keep the lineage of synapsids alive throughout the era of dinosaurs and eventually give rise to our own human line of descent. During the Mesozoic era, cynodonts would evolve into the group that we still call our own within the animal kingdom: mammals.

But so far in our story, cynodonts, together with the ubiquitous shovel-faced listrosaurs, remain a pitiful remnant of the synapsid lineage. Out of the ashes of the Great Dying emerges their rival clan, sauropsids, which would eventually include all the "saurs" of the Mesozoic—most famously, dinosaurs.

Mesozoic synapsid   Mesozoic sauropsid

Ugly Swans

In *The Last Day of Pompeii*, the volcano painting by Karl Bryullov that impressed me as a child, the Roman statues toppled by the eruption of Vesuvius are notably virgin white. This is how we moderns usually picture anything ancient Roman or Greek. Actually, classical art was quite colorful: it's just that paint has weathered away from all the ancient ruins.

It is the same with dinosaurs and with any other fossilized animal: there's a limit to what we know, beyond which all we have is imagination. Paleontologist Darren Naish and paleoartists John Conway and C. M. Kosemen bring this point home in their fantastic book titled *All Yesterdays*.[13] The first part of the book contains paintings and descriptions of dinosaurs as they *might* have been—not because there is any proof that protoceratopses could climb trees, duck-billed dinosaurs were chubby like piglets, or that plesiosaurs used mimicry to blend in with coral reefs, but because it is conceivable in principle based on the available fossils. The second part of the book, dubbed "All Tomorrows," imagines what scientists of the future would surmise about our current animals based on fossils—for example, they would have no way of knowing about rhinoceros horns, since those are made of keratin, like hair and nails rather than bone and would not be preserved in the fossil record. Based on skeleton alone, the scariest animal of our age is unquestionably the hippo, with its giant jaw and imposing tusks. A cow is a graceful antelope, and swans are repulsive creatures that feed by sticking their long, sharp arms into small prey.

The point is that when we try to picture animals of the past, we must use our imagination, but we also must be prepared to change the image if it turns out to be wrong. In fact, our idea of dinosaurs has already undergone a radical shift at least once.

The world first learned about dinosaurs in the middle of the nineteenth century. At that time, they were considered slow, clumsy, dumb, cold-blooded creatures. For example, B. W. Hawkins's life-sized models in London's Crystal Palace Park depict dinosaurs as big fat lizards on elephantine legs.[14] Even the word *dinosaur*, which means "terrible lizard," reflects this perception.

This Victorian image of the dinosaur changed abruptly during the second half of the twentieth century, in large part thanks to one dinosaur, the *Deinonychus*, and one person, the American paleontologist John Ostrom. Ostrom first saw a skeleton of a deinonychus in the early 1960s. It was about the size of an adult man. It had a thin, elegant frame that evoked speed and agility. Studying it made Ostrom question the classic image of dinosaurs as heavy, slow, and clumsy. This creature looked nothing like a sluggish lizard or a sleepy crocodile. It looked light, nimble, almost bouncy, ready to chase its prey at top speed, to sink its sharp teeth and big claws into pulsating flesh. This was not a "terrible lizard," Ostrom realized. This was a terrible *bird*.

Zoologists in the 1960s recognized that birds descended from some kind of reptile. But few saw them as direct descendants of dinosaurs per se. This was what Ostrom now advocated, citing an array of similarities between modern birds and fossilized dinosaurs. Birds weren't just related to dinosaurs or similar to dinosaurs, said Ostrom—they *were* dinosaurs.

At the time, the proposal was revolutionary. Today, it is considered an accepted truth. Over the past few decades, scientists have discovered that many dinosaurs were fast and agile,[15] colorful,[16] potentially smart,[17] perhaps even warm-blooded, just like birds.[18] Some took care of their young[19] and even provided for each other, which could be surmised based on the healing of severe wounds.[20] Some made nests,[21] some climbed trees,[22] some communicated with sounds—maybe even songs.[23] Most dinosaurs, according to the modern view, were not smooth and serpentlike, but sported plumage of various kinds.[24] Many had beaks.[25] There's no doubt today that dinosaurs have a lot more in common with birds than they have with snakes or lizards.

I often think about it when I see a bird. What Ostrom did was not just to upgrade dinosaurs to a faster, sleeker version of themselves, prompting a surge of scientific interest on the subject, termed the "Dinosaur Renaissance." He revoked the dinosaur extinction. Today, we call it the extinction of *nonavian* dinosaurs—meaning, *other than birds*. Dinosaurs did not die out. They remain a significant and in fact quite successful member of land fauna. This makes me look differently at eagles, crows, parrots, and starlings. I like to think that their world of air, light, color, and music gives us a glimpse into the Mesozoic, into a past that is forever lost but doubtlessly just as striking and infinitely diverse as these beautiful creatures.

Air Gills

Back in the Mesozoic, however, there was little to suggest a future in which a duck is considered an average-sized sauropsid. Almost from the onset of the era—meaning, after the Great Dying—"saurs" (dinosaurs were only the most prominent subgroup among a few others) soared to sizes that soon put even the most enormous animals of the Paleozoic to shame. An early example, *Fasolasuchus*, reached an impressive ten meters in length. No modern terrestrial animal is even close to that length—even anacondas rarely exceed five meters. *Brachiosaurus*, from the middle of the Mesozoic, reached the height of a five-story building and weighed as much as ninety horses. The most impressive example, to me, is from the very end of the era: *Quetzalcoatlus*, a flying pterosaur named for an Aztec feathered serpent god, was as tall as a giraffe, with a beak the size of a naval ram and the wingspan of a fighter jet—about ten meters. It might be the largest living creature that has ever flied. Pause to consider it—this was a real animal.

Sauropsids multiplied and diversified, taking over new habitats and niches, both carnivorous and herbivorous, as the world slowly recovered from the Great Dying. As sauropsids surged, synapsids declined. Listrosaurs, the "extraterrestrial pigs" that proliferated immediately after the Great Dying, waned over the next few million years. Cynodonts, our great-great-grandparents, continued to hold on to survival, but more and more branches were breaking off an initially diverse group—at the very beginning of the Mesozoic, racoon-sized cynodonts were not a rarity, but fifty million years later all that remained of them were tiny, shrew-sized creatures that all but disappeared from sight.[26]

We will return to them shortly but first let's take a moment to review the unlikely evolutionary comeback of their rivals, the sauropsids.

How did they manage to get the upper hand in the Mesozoic? Synapsids were the first to dominate, before the Great Dying reset the score, hitting the largest animals on top of the food chain the hardest and bringing everybody down to comparably small sizes. So at the beginning of the Mesozoic era, synapsids and sauropsids once again faced an even playing field, as they once already did. And this time, the outcome was the exact opposite of last time: synapsids retreated, whereas sauropsids boomed. What changed? The answer may have to do with oxygen—it would not be the first time. At the turn of the Mesozoic era, oxygen became a lot less abundant.

This was happening even before the volcanic eruption that triggered the Great Dying. By the end of the Permian period, oxygen levels already dropped by more than half as compared to the preceding Carboniferous, the lavish time of tropical forests, big amphibians, and dragonflies.[27]

The reason for this Permian oxygen crisis was the flip side of what made the Carboniferous so abundant. Surprisingly, the main cause in both cases was wood—a newly invented material that plants had recently deployed to reinforce their stems and become taller (not unlike the bones that allowed us vertebrates to become as large as we are). Today, natural equilibrium is maintained by decomposers—bacteria and fungi that digest the wood and return it to the atmosphere as carbon dioxide. But because wood was new during the Carboniferous, no decomposers had evolved enzymes to break it down yet. There were also lots of swamps where this wood was protected from slowly weathering away. No decomposition meant unused oxygen in the atmosphere—this explains why the Carboniferous was so oxygen rich and therefore animal-rich, too. But then, the rainforests collapsed, swamps dried out, and fungi caught up by evolving wood-digesting enzymes. The tide turned: remnants of undigested wood began to break down, consuming oxygen from the atmosphere and bringing it down to a lower, more natural level. (Not all wood ended up digested, though: a lot of it turned into coal and remained deposited in the ground, until humans came along and started digging it out to use as fuel—essentially completing what fungi couldn't finish.*)

So oxygen levels were already a severe pressure point for synapsids, whose oxygen consumption was prodigious compared to literally any other living creature that existed until that point. The Great Dying only exacerbated this problem, and that might have been what gave sauropsids a chance to take over.

One advantage that sauropsids have over synapsids relates to their lungs.

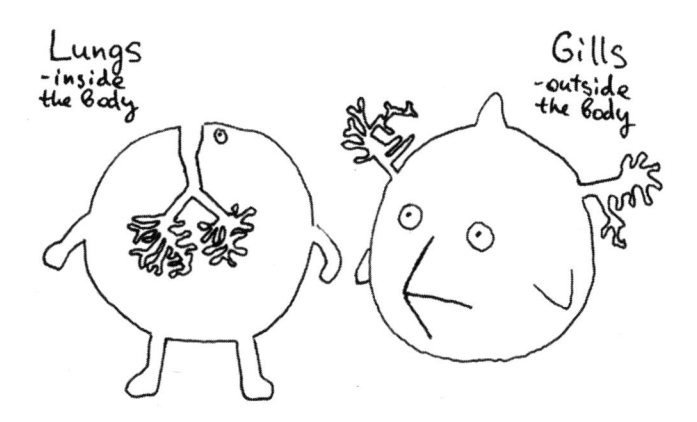

---

*Except we do it *much* faster than fungi ever could.

Synapsid lungs are essentially bags of air entangled in blood vessels. When we inhale, the blood absorbs oxygen from the lungs, and in a few seconds, we exhale oxygen-depleted air in the opposite direction. This reverse flow of air is totally useless in terms of oxygenating the blood any further: the only point of an exhalation is to make another inhalation.[28]

Bird lungs have a much more complicated design. Essentially their lungs are not bags but rather looped tubes circulating air in a continuous motion thanks to built-in air sacs that act like bellows. As a bird inhales and exhales, air progresses through its lungs gradually, entering at one end and exiting at another, so the air moves in one direction, rather than being pumped back and forth.[29] As a result, oxygen is absorbed not only during inhalations, but also during exhalations, nonstop. This advanced form of respiration is called unidirectional breathing.

Bird lungs combine the benefits of our more typical lungs and fish gills. The advantage of gills in the ocean is that they absorb oxygen continuously as the fish swims through the water, whereas lungs must be filled and emptied. This pumping is a waste of energy, and in the water, in which it is much harder to pump, the cost to do so would have been prohibitive. The problem with gills on dry land, however, is that they lose moisture too quickly: they are basically lungs turned inside out, and there's nothing shielding them from direct contact with the environment. If you want them to absorb oxygen, they will also evaporate water. There's another problem with gills on land: when you take them out of the water, they stick together, like wet hair after a shower. Oxygen can't suffuse their fine folds, and the gills become useless. This is the reason why fish suffocate outside of water (if you think about it, it's not obvious: there's a lot more oxygen in the air!). Lungs don't have this problem because they are inflated by the very process of breathing, which smooths out any folds—imagine inflating wet hair from within.*

The lungs of birds protect them from dehydration, as demanded by land life, but at the same time allow them to breathe continuously, almost like a fish. As a result of their unidirectional breathing, a bird can absorb significantly more

---

*Curiously, there are some exceptions from the overall rule that lungs are better suited for land and gills for water. Sea cucumbers have aquatic lungs (these creatures absorb oxygen with their entire body surface, so lungs just serve to extend it—there's no pumping), and coconut crabs have terrestrial gills (they prevent their gills from sticking together by making them more rigid).

Front air sac

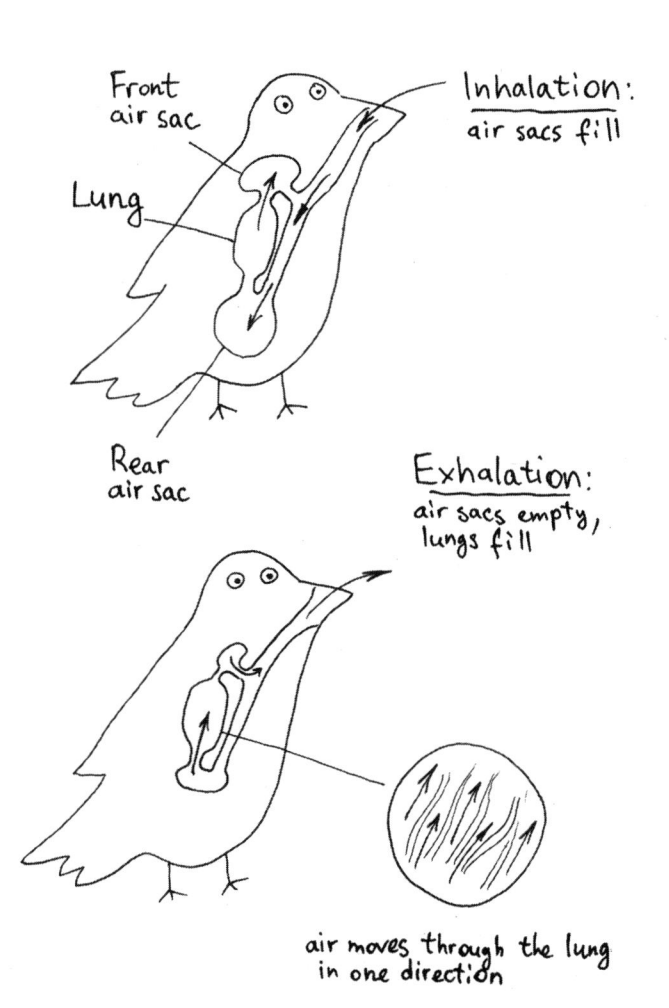

Inhalation: air sacs fill

Lung

Rear air sac

Exhalation: air sacs empty, lungs fill

air moves through the lung in one direction

oxygen per unit of time than a mammal of a comparable size. If you place a mouse and a sparrow into a hypobaric chamber that emulates oxygen-depleted air at six kilometers of elevation, the mouse will barely crawl on its stomach, whereas the sparrow will continue to easily fly around, showing no signs of distress.[30]

In the past, unidirectional breathing was considered a bird specialty, an adaptation to the energy demands of flight.[31] But more recently, it was also found in crocodiles, which clearly make no attempts to take off in flight.[32] On the evolutionary tree, crocodiles and birds represent two surviving branches that flank the extinct branches of all the famous Mesozoic dinosaurs. Of the species alive today, birds are the last surviving dinosaurs, and crocodiles are the closest thing to a dinosaur that's not quite a dinosaur. So if both crocodiles and birds have something unusual in common, extinct dinosaurs almost certainly had it too.

Evidently, unidirectional breathing is not just a bird thing, but also a dinosaur thing, and maybe even a sauropsid thing in general: it has since also been discovered in lizards, which relate to birds even more distantly than crocodiles do.[33] It is now clear that lungs with air sacs appeared long before birds learned to fly. So it's likely that the sauropsids that established the group's dominance

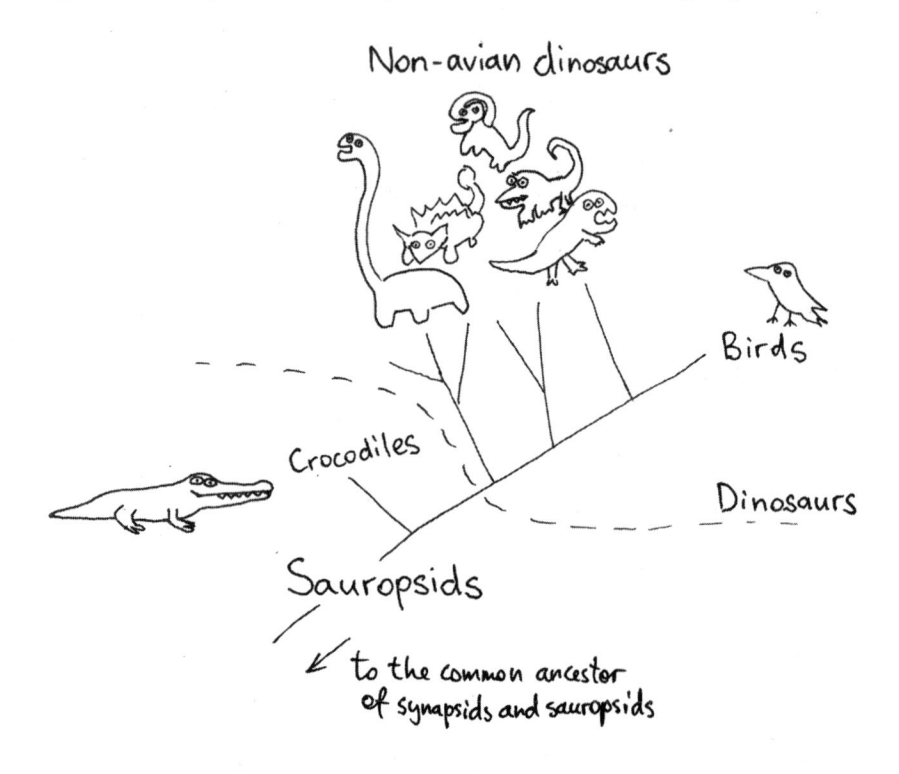

128 One Hand Clapping

in the early Mesozoic also possessed unidirectional breathing, which must have given them superior oxygen-absorbing abilities—and therefore, greater power and speed—compared to synapsids with their more primitive lungs. That seems like a perfect edge to have had at the time of catastrophic oxygen scarcity after the Great Dying.

There was another invention that may have helped speed up the sauropsid body: running. It appears that the ancestors of dinosaurs were the first runners in a world where everyone was still waddling around. This required some major restructuring of the sauropsid pelvis. Synapsids learned to run much later and did it in a different way, known as the gallop, in which the spine undulates up and down, rather than left and right as it does in running reptiles.[34] This, by the way, is the reason why whales and dolphins have tails in a horizontal orientation parallel to the plane of the water, whereas fish have tails oriented vertically. Ichthyosaurs, the reptile predecessors of dolphins, also had vertical tails, like fish. Both ichthyosaurs and dolphins evolved from terrestrial, four-legged vertebrates that at different points returned to the ocean and assumed a fishlike form. But ichthyosaurs were sauropsids, and their swimming developed from the ancient gait of their four-legged reptilian ancestors, itself dating back to fish and even earlier chordates, who all wiggled from side to side when moving—just like worms. Whales, however, are synapsids that only entered the water in the Cenozoic era, and so their mode of swimming developed not from this classic gait, but from the gallop, and so the motion is up and down rather than left and right.

It appears that during the Mesozoic, *speed* was the law of the land. Both breathing and running helped accelerate sauropsids beyond the more sluggish synapsids at the time when the latter were already struggling from an evolutionary equivalent of altitude sickness. This great acceleration is probably what secured the throne of the Mesozoic for sauropsids, as their former rivals entered a dark age. But as we are about to see, it is thanks to this very darkness that synapsids will soon catch up with the new, faster world of dinosaurs.

The Nocturnal Bottleneck

To us humans, sleeping during the night and walking around during the day seems natural, and indeed it is an obvious choice absent other considerations, since nighttime is both darker and colder than daytime. Dinosaurs were diurnal, and most of the large synapsids that preceded them in the Permian probably were too. Today's birds and reptiles are mostly active during the day as well. And yet mammals as a whole are a predominantly nocturnal group.[35] Humans and other daytime primates are more of an exception than the rule, and although there are other diurnal mammals today, they carry a distinct imprint of a nocturnal past. This is especially noticeable in their sense organs.[36] Compared to birds, most mammals are not very good at distinguishing colors and seeing far away, but they can see more at low light and are better at distinguishing smells, sounds, and textures.

All of this suggests that mammals, as a group, descend from nocturnal ancestors—small Mesozoic cynodonts that escaped the threat of dinosaurs by hiding during the day and only coming out when the dinosaurs were asleep.

This idea—basically, that our ancestors spent the Mesozoic in the dark—is termed the nocturnal bottleneck hypothesis. (Every semester I explain to students that a nocturnal bottleneck is not a body part that synapsids grew in the Mesozoic—someone always manages to assume so.) In fact the term refers to the narrowest of evolutionary paths that synapsids managed to crawl through during this era of sauropsid dominance. At the time, there were thousands of species of dinosaurs, and today, there is an equally vast diversity of mammals. The diversity of synapsids in the previous era, just prior to the Great Dying, was also comparable. But the diversity of cynodonts, which survived the Great Dying and the dinosaurs to eventually give rise to the mammals, was limited to some ten to twenty types barely distinguishable from one another—they were all small, insectivorous critters active at night or during twilight. For an unimaginably long period of time—the Mesozoic lasted 186 million years in total—our lineage barely held on to existence, and in another universe could have easily gone extinct. But it didn't: hence, a bottleneck.

When cynodonts were forced into this nocturnal exile, they struggled with two problems: low light and low temperature.

Low light meant difficulties with vision—the most effective sense for a daytime animal. Mammalian ancestors had to modify their eyes. Compared to other vertebrates, the retina of a typical mammal contains more rods, cells that are better at sensing low light but don't detect colors, and fewer cones, color-specific cells that require brighter illumination. There are also fewer types of these color-specific cones. Most vertebrates—fish, amphibia, reptiles, and birds—have four, whereas most mammals have only two, so from a bird perspective, mammals are all color-blind. The only sensible explanation for this is that mammals cleared space in their eyes for low-sensitivity vision based on rods rather than cones, as demanded by their nocturnal lifestyle.

Other sense organs were also modified—if your eyes aren't much help, you need to diversify. The middle and inner ear developed to include three ossicles and an elongated cochlea that resembles a snail. This made mammalian hearing more sensitive, especially at higher frequencies, which allows for more precise sound localization—such as when catching a chirping insect. Olfactory bulbs—the extensions of the brain that deal with smells—enlarged in size, and nasal passages became more complex, enhancing odor detection. Sense of touch also improved—although for most humans this sense is supplementary to vision and hearing, it is important for many mammals such as moles or shrews: they use their whiskers to feel out a narrow space and allocate a large part of their brain for processing this information. This is almost as bizarre to imagine as echolocation.

So, all in all, darkness forced mammalian ancestors to diversify their senses, which ended up including many independent ways to perceive the world at the same time and in great detail. As we see in the third part of the book, this pressure to sense the world in many different ways culminated in the advent of the cerebral cortex, the mammalian brain's *universal machine of understanding*, which consolidates all our different senses into a cohesive, conscious perception.

But besides the darkness, there was that second challenge of nocturnality—the cold. It is that second challenge that led our ancestors to their most unusual, counterintuitive, and arguably most important evolutionary innovation, which opened the path for the arrival of mammals.

The Speed of Everything

Prior to the colonization of land, it was practically impossible to maintain an organism at a temperature distinct from surrounding water. Water rapidly absorbs heat, and so it is very difficult for aquatic creatures to maintain body temperature higher than that of the water itself.

On land, however, an animal can start experimenting with body temperatures that exceed ambient temperature. Why would an animal want to be hotter than the environment dictates? The answer is that temperature controls many things in the organism—you could say it controls *all things*. Temperature is the speed at which molecules zoom around and ram into one another. (This ramming is what we experience as heat.) The speed of molecules determines the speed of chemical reactions. The speed of chemical reactions determines the speed of nutrient breakdown, the speed at which impulses travel along nerves, the speed of muscle contractions, and everything else that the organism does. So a hotter organism means a faster organism.

Living organisms generate heat simply by being alive. But for most animals, this amount of heat is so small that it immediately dissipates into the environment without any noticeable effect on the body.

Heat is lost through radiation or evaporation from the surface of the body. The larger the animal, the less surface area it has with respect to weight. So a small animal loses heat very quickly, but a large animal takes longer to cool down, just like a heavy cast iron pan takes longer to cool down than a light nonstick pan. For the same reason, two people can warm up by hugging each other, essentially becoming one double-sized animal and thus reducing their relative surface area. A large animal on land can amass substantial heat in its body simply by inertia.[37]

Prior to the Great Dying, synapsids weighed somewhere between twenty and one hundred kilograms[38] and so must have been at least partially warm-blooded in this passive way. One piece of evidence comes from the shape of their nasal passages, which tend to be more complex in warm-blooded animals—the extra surface area moisturizes the air and protects olfactory cells from drying out in the warm environment.[39]

Our line of descent runs through some of these large creatures and later through their much smaller, nocturnal successors weighing between twenty and one hundred *grams*.[40] At some point in their evolution, then, our ancestors were adapted to body heat, their enzymes, cells, and neural pathways optimized for a warm-blooded lifestyle, but then they found themselves restricted to the cold night and simultaneously reduced in size, meaning they could no longer hold on to their internal body heat. Whereas a large animal, in principle, does not need to do anything extraordinary to stay warm, this task is considerably harder for an animal weighing a thousand times less.

What were the options for the surviving synapsids? The most obvious one was to give up and stop trying. Lizards, for example, are also unable to retain much of their internal heat because of their small size. Yet they are quite good at warming themselves in the sunlight, and as a result of this continuous heat recharge their blood can be quite warm during their active period.* Lizards do run pretty fast, although they tire more easily than true warm bloods.[41]

But basking in sunlight was not an option for Mesozoic synapsids, whose only escape from dinosaurs was in the dark of the night. Had they really let go of their body heat, even a lizard lifestyle would have been out of reach. They would have had to settle for something even more pathetic: a slow, hazy, docile existence among night frogs and other evolutionary has-beens.

There was, however, another option, one never attempted before by anyone.

Jailbreaking Metabolism

Thanks to gulags and decades of mass incarceration, which directly or indirectly affected most families in the Soviet Union, prison lore forms a significant undercurrent of Russian culture. When I was growing up, every teenager knew the recipe for making *chifir*—a boiled-down black tea so strong that it becomes intoxicating: you put a tin of loose black tea into a bucket of water, then cut a high-voltage wire, and stick the two bare ends into the bucket. No one I know had actually done this and neither should you, but supposedly the water boils in minutes, and prisoners have a good time.

The warm-bloodedness of mammals follows a similar logic. Previously, any heat generated by the body was a byproduct of some useful process, like digestion or muscle movement. But mammals go beyond that: they essentially break their own metabolism in such a way that it wastes copious amounts of energy. They cut the wire and stick it into the bucket—just so they can burn more calories and thus produce more heat.[42]

---

*This example illustrates why the terms "warm-blooded" and "cold-blooded" are considered too imprecise by modern zoologists—I am using them for the sake of simplicity. It is more informative to describe lizards as poikilothermic ectotherms, meaning creatures with an unstable body temperature controlled by external conditions. Mammals, by contrast, are homeothermic endotherms, meaning animals with a stable body temperature determined by internal processes.

Heat is the form energy assumes by default—you could say it is energy's favorite state, a place where it always wants to go. So, typically, a living organism does anything it can to *prevent* energy from turning into heat and instead to divert it toward useful action.

The key element of the cell that harvests energy from nutrients is mitochondria. Mitochondria take in the energy-rich molecules we consume, break them down atom by atom using oxygen, and as their final product, churn out ATP, a molecule that serves as a standardized energy currency of the cell into which all nutrients are converted. But there's an intermediate step in the process, during which the energy collected from nutrients is converted into an unusual form: not yet another chemical, but pressure. Just as it is possible to convert the chemicals in burning fuel into the pressure of steam in a boiler, it is possible to convert chemical energy of nutrients into pressure inside mitochondria—except in this case the pressure is generated not by steam, but by protons (free-floating hydrogen atoms that are missing their single electron—there's always a lot of them hanging around in any water solution, and it's not critically important that it's protons, specifically), and the pressure builds up not in a boiler, but in the narrow space between mitochondria's two membranes. So, first, nutrient energy is used to pump a lot of protons into that space, and then these pressurized protons are released through a molecular turbine called ATP synthase, just like steam in a steam engine. ATP synthase is not just some metaphorical turbine—it is fully literal, complete with a rotating hexagonal rotor inserted into a static part anchored in the mitochondrial membrane. As it rotates, it churns out the all-important ATP, which is then used by all other molecules of the cell to perform all their useful functions.

So to successfully convert food into ATP (and thus into cellular function), pressurized protons must pass directly through the ATP synthase. Mammals, however, add a twist to the process. Occasionally, they insert into their mitochondrial membranes, right next to the ATP synthase, a protein called UCP1. This UCP1 is a pore through which protons can exit their pressurized space. But in contrast to the ATP synthase, UCP1 does not make any ATP—or anything useful, for that matter. It simply lets protons leak out without spinning any turbines, which is, ostensibly, the whole point of the process.[43] What happens to all the energy that would otherwise be collected? It is lost. Meaning, it is dissipated as heat.

This is about as crazy as sailors warming themselves up by making holes in the hull of the ship. The mitochondrion simply wastes energy by idling—and the more UCP1 it contains, the more it idles. Energy passes from nutrients straight to heat. The "idlest" UCP1-rich mitochondria are found packed into the cells of a characteristically mammalian tissue named brown fat. "Fat" refers to these cells' reserves of this energy-rich nutrient, whereas "brown" refers to their abundance of mitochondria.

A more typical fat tissue, or white fat, contains far fewer mitochondria and is focused on long-term storage of fat, not its constant consumption. The essence of white fat is storing the fat for a rainy day; the essence of brown fat is continuously burning it for heat. (For this reason, a lot of medical attention is focused on ways to convert white fat into brown fat—call that "Mesozoic weight loss.")

The idling of mitochondria is not the only process that wastes energy for no reason other than producing heat. Other examples include the "idle" shuttling of sodium and potassium ions in and out of the cell (more on that later)[44] and "idle" muscle contractions—shivering. These are not discordant processes but a unified set, as evidenced by the fact that the many variations of such metabolic idling are simultaneously controlled by thyroid hormones. The thyroid gland basically acts as the master regulator of the body's energy wasting program.[45]

To understand why this is so unusual, we have to suspend our human bias. To modern humans, having a body that wastes its own energy sounds like a great idea, because for the majority of our species, excess fat far outweighs starvation as a health risk. The average human has access to foods more energy-dense than any edible material planet Earth had ever seen—like French fries. The prime concern across our species is what to do with all this nutritional energy so it does not overload our bodies with fat deposits.

But this is not how it generally works in nature. If there is a lot of food, soon there will be a lot of consumers and thus no longer a lot of food. In the long run, energy is always scarce. So a mammalian body, on the evolutionary gambling table, is an all-in bet. A typical mammal body consumes about five times as much energy as a typical reptile of a similar size.[46] It is not just an incremental difference—it's a whole new idea of how to approach energy use. What is so novel about it is that we mammals spend a colossal amount of energy not as a byproduct of something else our body does, but for no reason other than becoming ridiculously, unnaturally hot. It is an extremely dicey gambit, but it makes us the fastest animals on the planet. We are like race cars built with total disregard for fuel use because the only purpose is to extract maximum power from the engine.

Many human-made devices, like a CPU of a computer, require cooling rather than heating. This is because for most CPUs, the energy supply provided by the power grid is virtually unlimited, so the problem is not preserving the energy, but making sure all the energy use does not melt the device. Most animals never have to deal with anything similar because their metabolism is not intense enough to cause overheating. With the advent of warm-bloodedness, however, heat becomes an evolutionary goal in and of itself, and so body temperatures soon rise to the point at which they become hazardous, just as overheating is for CPUs. We operate very close to the limit of safe temperatures. Many of our cells and molecules cannot stand even a few degrees above their normal set point. If our bodies uniformly warmed by 10 degrees Celsius—the equivalent of a lukewarm beer brought to room temperature—our blood would clot. Since the great acceleration of the Mesozoic era, our bodies are constantly ramped up to the edge of biochemical possibility.

The "nocturnal bottleneck" began as a forced exile into oblivion. But eventually, it set the stage for the most unlikely resurgence of the ancient synapsid clan. Dinosaurs sure left a mark on our ancestors—a mark that we still carry in our bodies, in our eyes, ears, and brains. But it is that very mark that would distinguish our lineage in the years to come. Even as dinosaurs were still establishing their dominance, seeds were already sown for the future triumph of their successors—mammals, also known by another old-fashioned but simple, beautiful, and ancient name, *beasts*. It was them—*us*—who were destined to rule the world when dinosaurs finally went extinct. What remained was to wait for the asteroid.

Between Two Ends

The Mesozoic began in Siberia but ended on the Yucatan. This peninsula in Mexico is famous for its cenotes, sinkholes opening into deep underground wells, where pools of turquoise blue water are illuminated by the tropical sun beaming down in an ethereal glow.

Mayans considered cenotes sacred portals to the world of the dead, which is frankly a bit of a trope, but in this case there's really something to it.

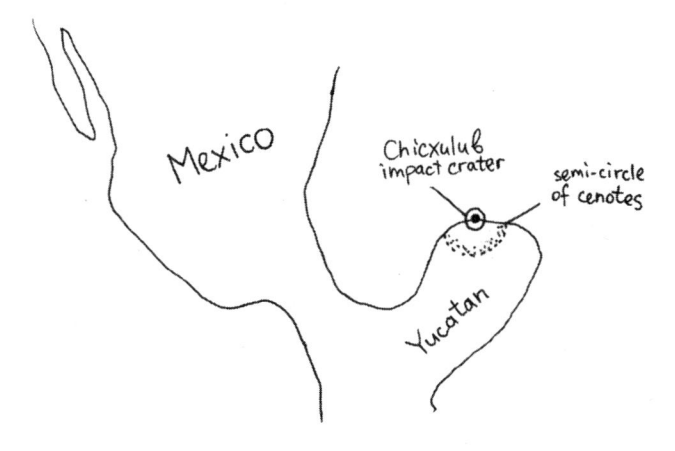

On a map of Yucatan, the cenotes form a semicircle of about 180 kilometers in diameter. If the semicircle were made into a circle, its center would fall on a point a few kilometers off the coast of the peninsula. The reason for their circular arrangement became clear only in the last few decades, with the most definitive confirmation coming in 2016. Scientists drilled into the center of the circle—the one just off the coast of Yucatan—and found molten rocks produced by one of the largest asteroid impacts in the history of planet Earth.[47] The semicircle of sinkholes, as it turned out, marks the boundary of the vast crater the asteroid formed upon collision, known as the Chicxulub crater. The name is derived from the locality and translates from Yucatec Mayan as "devil's flea"—a great descriptor for the asteroid that bit through the skin of the planet sixty-six million years ago, triggering yet another mass extinction and another change of eras.

It was almost a reenactment of the Great Dying. There was, once again, a global fire, still visible in the geological record as a thin black line separating Mesozoic and Cenozoic deposits; acid rain, caused by sulfur from gypsum deposits vaporized by the impact; and an "impact winter"—prolonged climate change caused by the dimming of the sky. Once again, the world faced a cataclysmic event. Once again, it wiped out the largest, most energy-demanding, most biologically ambitious creatures there were. Only now it was dinosaurs—all but birds—who found themselves in this unfortunate position. Their extinction marks the beginning of a new era that is still ongoing today: the Cenozoic.

For the *third* time in history, the great game of synapsids and sauropsids was back to square one—the top of the food chain, once again, stood vacant.

Except synapsids, by now, were unrivaled experts in the game. They had been living a postapocalyptic life for an entire era, so even though the Chicxulub impact reduced their numbers, too, they quickly shrugged off the added duress and began expanding again. It had been a perfect case of "what doesn't kill you makes you stronger"—*and the closer you come to dying*, mammals might have added, *the stronger you become*. Ancestral mammals stood at the brink of extinction for a very long time and were forged by these years of hardship into the fastest and smartest creatures the world had ever seen. It was time for the most stunning evolutionary reversal in the history of the planet—a true royal restoration.

With the big dinosaurs gone, a pressure valve was released on mammalian life. While the world was still recovering from the Chicxulub impact, mammals began to take over the niches that used to belong to dinosaurs. In some ten million years (compare that to 150 million years of the nocturnal bottleneck), mammals were at the top anywhere there was animal life, diversifying and adapting to a myriad of novel lifestyles. Each of today's subgroups of mammals is a realization of some new possibility that opened in the Cenozoic era. Carnivorans (dogs, cats, bears) are expert hunters. Ungulates (gazelles, antelopes, horses) are expert runners. Cetaceans (whales and dolphins) represent mammals in the sea; moles, underground; and bats, in the sky, although mostly during the night. (We consider the specialty of our own subgroup, primates, in the next chapter.)

It was a triumph. In a blink of evolutionary time, mammals leapt from the shadows into the spotlight of planet Earth's megafauna, replacing reptilian rivals that nearly cost them their existence. In this spotlight, mammals remain to this day.

The Eternal Rivals

And what of the dinosaurs? Just as synapsids did not go fully extinct in the Mesozoic, neither did dinosaurs in the Cenozoic. It's just that out of all the multitude of the Mesozoic monsters, only one branch survived—a group of small, stealthy creatures that made an evolutionary bet on avoiding an enemy they could not compete with. Our ancestors made the same bet an era ago—they escaped from dinosaurs into the night. Birds, the surviving dinosaurs of the Cenozoic, escaped from mammals into the sky. Daytime flight remains the stronghold of sauropsids—it might be the only vertebrate lifestyle that mammals avoid with rare exceptions.*

And so the great rivalry goes on.

What is most incredible in this synapsid-sauropsid dichotomy is how similar the two clans remain for hundreds of millions of years, despite how ancient their evolutionary split is and how many rounds their great game has already seen. Typically, big evolutionary victories are permanent. Insects, for example, for purely physical reasons never had the chance to come back and regain their place on top of the terrestrial food chain once vertebrates entered the scene. It is equally difficult to imagine amphibians suddenly rising back to prominence and outcompeting amniotes. But sauropsids and synapsids just keep coming

---

*However, considering that airplanes move about four billion people per year—that is, more than half the biomass of one the most populous species of animals on the planet—the most "aerial" of the vertebrates are humans.

back to face each other again and again, their stories unfolding as a deathly dance in magnificent swings over three geological eras.

The extent to which our two clans continually parrot each other's evolutionary inventions is almost unbelievable. This is called convergent evolution, and I can't think of any two evolutionary branches in the history of life on Earth that do so much of it over such a long period of time. Take for example the skull. The synapsid opening in their skull originally helped them with jaw muscle attachment. As far as we can tell from fossil evidence, this trait is what set them apart from sauropsids in the first place. But it took only a moment of evolutionary time for sauropsids to start catching up: it is as if they realized they had been missing out on a great invention and started adding openings and arches to their skulls as well. You could argue that sauropsids took the idea even further than synapsids: today's snakes have almost wire-frame skulls that are literally all arches and openings, whereas in humans, the original synapsid opening can take a minute to even find.

Another example of convergent evolution between synapsids and sauropsids is the structure of the heart. Amphibians, such as frogs, have a three-chambered heart: two atriums (chambers for incoming blood) but only one ventricle (chamber for outgoing blood), in which blood from both atriums mixes before getting pumped into the body. Both synapsids and sauropsids independently developed four-chambered hearts, in which the single ventricle was split in two, separating arterial and venous blood and so making oxygen distribution more efficient.

The list goes on. Running? Sauropsids did that first, and synapsids caught up later with their gallop. Ossicles in the middle ear? Synapsids developed three of them and so, eventually, did sauropsids, although mammals still hear better at high frequencies. Warm-bloodedness? The body-accelerating "energy wasting program" was a mammalian novelty early in the Mesozoic, but in the Cenozoic, birds managed to achieve even higher body temperatures, although by slightly different means. Synapsids and sauropsids are always breathing down each other's necks, pushing each other to evolve, innovate, and never give up.

Like a great sparring partner, we have a lot to thank the sauropsid clan for, even though our ancestors struggling through the nocturnal bottleneck would have begged to differ. It is true that we really dodged a bullet

Amphibian heart

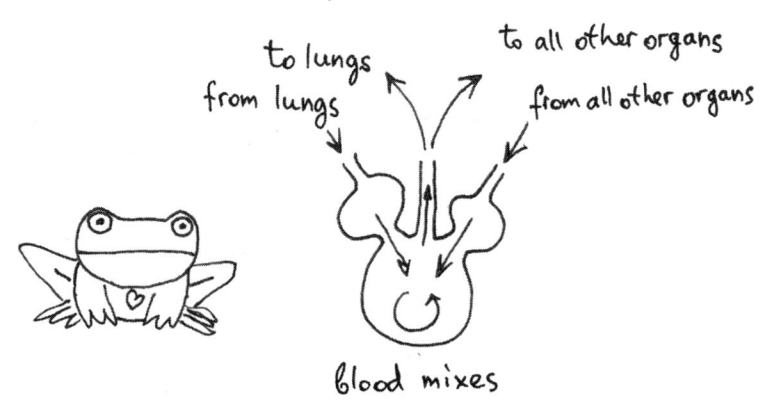

to lungs
from lungs
to all other organs
from all other organs

blood mixes

Mammalian heart

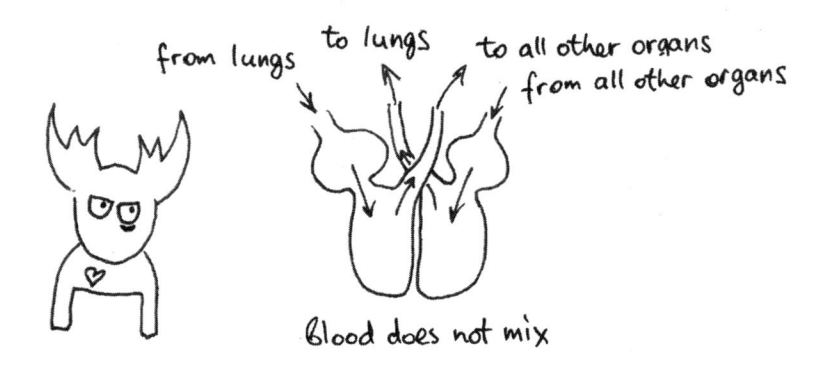

from lungs
to lungs
to all other organs
from all other organs

blood does not mix

in the Mesozoic, but without that era we may never have become the kind of intelligent and lightning-fast, yet gentle, social, and playful creatures that humans—and all mammals—are. In the next chapter, we see how this last part—the mammalian social instinct—is also rooted in the dark days of our ancestors' struggle against dinosaurs.

Today, if you compare the total biomass of all land vertebrates, you'll find mammals far ahead of birds by a factor of roughly 22 to 1. But that balance is overwhelmingly tilted by humans and livestock. If you consider only wild birds and mammals, the ratio is much closer: about 3.5 to 1. Certainly by now, we should know better than to declare the great game to be over. Watch out for birds—they might still have a few cards up their sleeve for the next apocalypse.

Chapter 8.
The Mirror

Long after her death I felt her thoughts floating through mine. Long before we met we had had the same dreams.

VLADIMIR NABOKOV

One of the most transformative pieces of documentary footage I've ever seen is a segment of BBC's *Seven Worlds, One Planet*, directed by Abigail Lees and centered on an albatross nest on an Antarctic island. Albatrosses are stunning: they have the largest wingspan of any living bird; they fly around the world, soaring over entire oceans, using their extremely developed senses of smell and sight for navigation; they have prolonged, elaborate courtship, including charming synchronized dances, and they tend to pair bond for life. In the documentary, an albatross chick is waiting for its parent to return to the nest, a small stump on the ground made of mud, sticks, and feathers. A storm breaks out, and the wind blows the chick out of the nest. The little bird tries to climb back in, but it is weak, wet, and cold, and although the nest is only about as tall as a teacup, the chick just can't make it over the edge. Finally, the parent returns, and viewers breathe a sigh of relief. The graceful giant lands on its nest, sits down comfortably, and proceeds to *stare blankly at its struggling, dying baby*. Albatrosses, we are told, don't recognize their chicks by sight, sound, or smell, but merely by the fact that they are located within their nest. If the chick is not in the nest, the parent simply doesn't register it as something to pay attention to. It actually blocks the chick from climbing into the nest with its own body. Somehow, the chick finally manages to muster energy for a decisive push, and the parent happens to stand up at this very moment. The chick plops back into the nest between the parent's legs, and only then, like a switch, the parent suddenly sees it. It bursts into action and for the next two days feeds and warms the chick, showing nothing but exemplary parenting.

There are many other examples of birds behaving in such eerily robotic, rigid ways that seem stupefying from a human perspective. Ducklings can get imprinted onto a random object, like a ball, which they then follow around instead of their mother.[1] Songbirds continue to feed a parasitic cuckoo chick even when it's twice their size after the intruder pushes their own chicks out of the nest.[2] An ornithologist friend told me she was once trying to label baby tree swallows by painting their nails but found that parents threw chicks with white nails out of the nest, presumably identifying anything white as feces.

None of these are dumb animals. Albatrosses may not be as intelligent as crows, but they are advanced enough to have distinct personalities, to navigate over thousands of kilometers, and to return to the same nest for decades.[3] They seem, at least at the surface, to be at the intellectual level of an average mammal—maybe a dog, if not a monkey. But you could *never* imagine a dog, a monkey, or even a rat failing to recognize their own offspring desperately crying right in front of them, simply because it doesn't check some mental box that says "my offspring." That is simply not how our mammalian brains work.

So although parental care is yet another parallel between birds and mammals, adding to the many similarities discussed in the last chapter, it means something different for our two groups. Birds, generally speaking, seem to perceive their chicks as objects with special status and become easily confused about which objects qualify. Mammals, however, base their care on an idea that appears unfamiliar to an albatross: *this being is the same as I am.*

This idea, as we are about to see, is the foundation for the entire mammalian social instinct—not just the ability of parents to recognize their babies, but our general ability to understand each other's actions, feelings, and thoughts.

I Feel You

Primatologist Frans de Waal would show in his lectures a video featuring two capuchin monkeys housed in separate, side-by-side cages. A scientist has them perform a task for her: each monkey must hand her a rock from the cage and receive a reward in exchange. But they receive different rewards: the first one gets cucumber (which the monkeys are perfectly fine with as long as there's nothing else on offer), whereas the other gets a grape (a far superior food among capuchins). Initially, the first monkey eats the cucumber and seems content with it. But when it sees the second monkey receive a better reward for the same task, it erupts in anger, throws the cucumber back at the scientist, tries to grab food out of her hands, then darts back to the cage and taps another rock against the wall before handing it over ("Is there something wrong with my rocks?"), but gets cucumber once again. The video ends with the monkey pounding the cage walls and howling furiously at the injustice.[4] (You can find this video by searching "capuchin monkeys reject unequal pay.")

The thing is, capuchins are not anti-cucumber. If it weren't for the second monkey receiving a grape, the first monkey would quite enjoy being paid in cucumber. The rage is not about the food itself—it is about comparing one's own experience to someone else's. The capuchin imagines itself in its neighbor's shoes, and by performing this mental operation realizes that its own shoes are not as good.

This is how we understand other people, too, in the most general sense. When we listen to someone's speech, we repeat the words in our head as if we were the ones formulating them. This is best seen in people who have difficulties with understanding, such as those with Wernicke's aphasia, a neurological condition that affects the link between words and their meaning. Patients with this type of aphasia (typically the result of a stroke) often try to repeat the other person's words as quickly as possible to grasp their meaning—and when they succeed in this replay, they also succeed in understanding, but the words often crumble, turning into gibberish, in which case their meaning eludes the patient.[5] When we look at a person experiencing pain, we experience a part of their pain as if it were our own—that can be seen in the activation of regions in the brain responsible for actual pain. When we look at someone else's emotions, we get infected with them, as if they were ours to begin with.

We don't just react to other people—we *try them on*.

It's not just humans and monkeys. It's mammals in general. Even rats do this: if you trap one rat in a cage and let another rat—a total stranger—see that, the second rat will set the first one free and even share with it a piece of chocolate it would have otherwise eaten itself. It's hard to explain why evolution would allow such seemingly selfless behavior—wouldn't egoistic rats end

Egoistic rat eats all the chocolate

Altruistic rat gives the chocolate away

up outcompeting the altruists over time?—unless you allow that the essence of this behavior is not *selflessness* per se, but rather *seeing another rat as itself*, which, despite the split chocolate bar, has many other evolutionary benefits.

How does the brain handle such a complicated and multifaceted mental task, comparing someone else to oneself? In the 1990s, some brain scientists thought the key to this ability—and with it, to empathy and sociality—was to be found in the so-called mirror neurons.[6] These neurons become activated both by performing an action *and* by watching someone else perform it. For example, when a monkey reaches for a toy on the table, a particular set of neurons "lights up" in its motor cortex. Then the monkey watches a scientist reach for the same toy—the same set of neurons is activated, as if "reflecting" the actions of another being, hence "mirror neurons." It was eventually shown that other species, including rats and humans, also have neurons that behave like that. In human subjects, if these mirror neurons are temporarily turned off (this can be done using a safe magnetic pulse, no surgery needed), the person has trouble interpreting others' actions.[7]

Today, mirror neurons are considered only one expression of a broader principle that underpins sociality in mammals: modeling other beings using our own brain.[8] Neurons in the motor cortex model others' movements—that's the original "mirror" that scientists were able to find, and so the phrase "mirror neurons" became associated with them. But in the same way, neurons in the emotional cortex model others' emotions, neurons in the language cortex model others' words, neurons in the pain cortex model others' pain, and so on. The key to our social intelligence is that our entire brain acts as one big mirror.

No mammals use this mirror as skillfully as primates, and among them no species has mastered it as well as Homo sapiens. But before we turn to what makes monkeys, apes, and humans such uniquely social creatures, let's consider mammals as a whole.

A Billion Years of Solitude

Most members of the animal kingdom spend their adult life in solitude and only think about company when it is time to reproduce. Sometimes even that is not required. A fish, for example, could theoretically hatch from an egg, grow up, lay its own eggs, and die without ever meeting another fish from the same species (though this is unlikely in practice). But mammals have at least one guaranteed moment of interaction with a related being: live birth.

Based on fossil evidence, it appears that synapsid mothers were actively involved in rearing their young even prior to the dinosaur era. The earliest known evidence for maternal care among synapsids dates as far back as the Paleozoic: a mother and four of her young.[9] Cynodonts, the synapsids that eventually gave rise to mammals, have also been found in similar family units.[10] There was, however, an important difference distinguishing these ancient synapsids from modern mammals: they were oviparous, meaning they laid eggs.

Back in the Paleozoic, watertight amniotic eggs were the height of evolutionary technology. It was thanks to them that amniotes walked over amphibians to achieve land hegemony. But moisture and dehydration—the problems that amniotic eggs solved—were a mid-Paleozoic obsession. By the end of the era, evolutionary priorities had shifted to oxygen, and during the harsh Mesozoic, the world seemed to turn upon temperature. In this new reality, amniotic eggs became a burden.

Forced into a nocturnal existence and diminished in size, the ancestors of mammals ran into the problem of excessive heat loss, which, as we saw in the last chapter, they solved by becoming warm-blooded. But this still left their eggs, which were also reduced in size. As a result, they also lost heat at a higher rate, which the embryos could not possibly counteract.

It was not enough for warm-blooded adults to warm themselves—they also had to provide heat for their offspring. This could be done in two ways.

One is to have the adult sit on top of the eggs. Another is to place them inside the adult's body.[11]

The first option is employed by today's birds, which cannot afford any extra weight because of flight. Instead, they make huge eggs (large enough to have a substantial reserve of heat and nutrients), put them in a nest, and sit on them. This is risky, because the eggs have to sometimes be left unsupervised, so they can lose all the precious heat or, worse, be eaten—bird eggs are nature's favorite fast food. But birds tend to fly around during the day when it is warmer and typically leave their eggs in places where nothing can reach them. So, for birds, large stationary eggs are an acceptable option.

For early mammals, that wouldn't have worked as well, since they lived on the ground and were active during the night, so their eggs would have faced a much higher risk of predation and excessive cooling. Most mammals, therefore, took the other route: rather than trying to maximize heat capacity and nutrient storage, they reduced their eggs to the minimum possible size and simply connected them to the maternal organism from the inside for the entire duration of development.

Development inside the mother's womb, or viviparity, gives mammals plenty of time and the most perfect environment to construct their organisms. This greatly expands their capabilities in adulthood. But there's another, less obvious consequence of viviparity: it guarantees that we meet our mother.

This meeting is where all mammalian relationships begin. There are good reasons to believe that the rest of our relationships—with family, friends, partners—are based upon the mental glue that binds together mothers and children.[12]

Case in point is a molecule called oxytocin. It receives a lot of media attention as a "love hormone," which gives you a general sense of what it does, but it is actually a lot more nuanced and interesting. Oxytocin is a tiny, nine-amino-acid chunk of protein that spreads through the body and brain and alters the way cells work. Molecules similar to oxytocin have existed since the dawn of the animal kingdom, employed by both invertebrates and vertebrates for hundreds of millions of years as parenting hormones—signals regulating their reproduction.* This ancient role of oxytocin is preserved all the way to humans.

---

*Actually, the range of functions is broader and very surprising. It appears that in all modern animals, both vertebrates and invertebrates, oxytocin-like hormones regulate three things: brain, reproduction, and water excretion (for example, kidney function)—a puzzling combination, which, to my knowledge, no evolution-

One of the key effects of oxytocin is to stimulate uterine contractions during labor, at the same time reducing pain, so it literally helps women give birth. This, in itself, is remarkable, given that almost the exact same molecule also helps beetles lay their eggs.

But in humans oxytocin also works at a higher cognitive level. In addition to its effects on the body, it is also associated with psychological effects such as bonding and trust. During and after labor, both the mother and her child get a massive rush of oxytocin, which ensures that they will both be attached to each other.

But it is not the kind of attachment that albatrosses have to their chicks—an apparently mechanical reflex that makes them feed an object in their nest. Oxytocin does not simply turn on love (although it does actually have some pharmacological overlap with the party drug MDMA[13]). Instead, oxytocin boosts the brain's capacity for reflection.[14]

This doesn't just happen in mothers and children. Oxytocin can be released in the brain in response to a positive interaction with any other human, whether friendly, familial, or romantic: heartfelt conversation, prolonged eye contact, hugging, sex. And when it does get released, it increases generosity,[15] trust,[16] and mutual understanding;[17] boosts the ability to discern emotions through facial expressions;[18] and lowers the fear of making eye contact.[19] So prolonged eye contact makes you want to want make even more eye contact, and so on—it's a positive feedback loop, a self-reinforcing cycle of bonding that underpins friendship and love. And what's most elegant about this oxytocin system is that it's guaranteed to simultaneously engage both people, because all the triggers for oxytocin release—like eye contact, touch, conversation, even childbirth—are also simultaneously happening to both people. It is as if oxytocin was designed to increase the mental symmetry between two creatures. *Oxytocin is a response to reciprocity that causes reciprocity.*

Under the hormone's influence, we let people into our soul, feel out their personalities using our own emotions, and think their thoughts using our own words. The physiological connection of oxytocin to reproduction suggests that mothers were the first to do this. At some point in the evolution

---

ary biologist is quite sure what to do with. Presumably, there had been some ancient connection between water excretion and reproduction, which later came to additionally involve the brain.

of our ancestors, maternal reflexes—such as those of an albatross—gave way to maternal reflection.[20] Rather than treating their babies as mere objects to act upon, mothers began to see them as personalities, understand their needs, and predict their behavior.

But then, what started as maternal reflection became generalized. The ability to understand others and form bonds with them turned out to be a useful innovation that extended far beyond childcare. It represented a new form of interaction between two organisms, the utility of which soon outgrew the limits of the mother-infant bond. Essentially, mammalian sociality is a form of maternal instinct extended to other individuals, motherhood-plus.

It would not be fair to birds to conclude that they are wholly incapable of empathy or of putting themselves in someone else's position. Rooks, for example, are known to console each other when stressed,[21] whereas scrub-jays can keep track of which other birds have observed them caching food, indicating they can take the visual perspective of others.[22] But as a whole, mammals seem a whole lot nicer to each other—consider, for example, chickens, who sometimes peck their own chicks to death and drink their blood, and actually get encouraged to do in social settings. This kind of thing would shock even the most brutal mammal.[23]

Being a bird sure sounds fun—but I am grateful to be part of our kinder, gentler mammalian world.

Life Finds a Way

When non-avian dinosaurs died out, mammals rushed to fill the space they left empty. Like water that assumes the shape of a vessel, mammals were shaped by the possibilities that opened up before them. Dolphins assumed the shape of fish, cats grew sharp fangs and claws, and bats, membranes between their fingers. Moles evolved digging claws, deer, running hooves, and rabbits, massive ears that could detect a predator from miles away. Each subgroup became a specialist in something of its own.

What is the specialty of our own subgroup—primates?

When we look at monkeys or apes, I think we are too distracted by their similarity to ourselves to realize what the essence of a primate really is. It is, however, completely obvious once you point it out. Primate specialty is *living on trees*.[24] We are among the very few that don't.

Mammals

Of course, tree climbing is not exclusive to primates—think of squirrels, for example. But no one does it as well as a monkey. Based on fossil data, primates began with something pretty squirrel-like, except with elongated, flexible limbs and opposable first digits, much better suited for grasping tree branches than rodent paws are. As primates became further specialized for their arboreal lifestyle, this grasping ability developed into a unique mode of three-dimensional motion: leaping.

A leap is different from a jump. A jump, as it is typically understood, is made by the legs pushing the body against a surface. This is how squirrels move between trees. But a monkey's leap is made by swinging the weight of the body on a branch using the front limbs. It takes advantage of the body's weight and gravity to produce acceleration, rather than relying on muscle propulsion alone, so a monkey can travel between trees a lot faster and waste a lot less energy. Another advantage is that a leaping monkey faces the direction of its motion, whereas a jumping squirrel, not necessarily. As a way of moving in three dimensions, leaping is reminiscent of flying, except it is limited to the canopy.

There is only one problem with this tree-flying and three-dimensional movement through dense vegetation more generally: you have to see it. The different branches all smell the same and make the same sound. They are too far apart to feel them out with whiskers. So the only option is vision. And vision works a lot better during the day.

This was probably the main reason most primates left the nocturnal existence of the Mesozoic behind and transitioned back to a daytime lifestyle.

The order of primates first appears in the geological record about fifty-five million years ago, ten million years after the Chicxulub impact exterminated the Mesozoic dinosaurs. The last common ancestor of primates was likely a nocturnal and solitary creature, just like most mammals at the time. A handful of its descendants retained these features—these are some of our most distant primate relatives, such as lemurs and galago. But the vast majority of primates made two transitions, which, as we will see, are directly related to one another: they transitioned from nighttime to daytime and from a solitary lifestyle to living in groups.

In the previous chapter, we discussed the imprint of a nocturnal past even in mammals that are active during daytime: for example, all mammals see fewer colors than birds, but they are better at seeing in the dark. Primates have features that point to their mass transition in the reverse direction.[25] We see more colors than other mammals. This is explained by an additional color-sensitive cone in our retina—birds have four, dogs have two, and we have three—so we are still not as good as sauropsids at daytime vision but at least better than our nocturnal ancestors. You can tell this just by looking at monkeys and comparing them to other mammals. There's no point being colorful if other members of your species can't tell the difference.* So most mammals are brown or gray, in contrast to colorful birds—but monkeys do sit somewhere in the middle, often deploying vivid reds, yellows, and blues to supplement the dull mammalian palette.

Primate vision has another advantage: our eyes are positioned in the front of the head, which means both eyes always look at the same thing simultaneously. Squirrel eyes are positioned differently, on both sides of its head. Eyes

---

*Tigers, with their striking orange-and-black coloration, might seem like an exception. But if you are color-blind like most tiger prey is, tiger fur actually blends into the environment very well, providing excellent camouflage.

on the sides of the head are characteristic of prey, whose main concern is to detect danger coming at them from any direction—this enables them to see even behind themselves. By contrast, typical predators, such as cats or owls, tend to have frontal eyes, just as humans do. That narrows the angle of vision, but increases the overlap between two eyes, which means that almost the entire field of view is captured from two slightly different angles—and that allows the brain to compute depth, meaning how far away different objects are. It may seem counterintuitive that we need two eyes to see depth, because when we close one eye, we don't immediately notice that our vision goes "flat." This is simply due to habit. We spend most of the time using both eyes together, so we memorize the dimensions of objects. If, however, you cover one eye with a pirate eye patch and try walking around somewhere new, you are guaranteed to eventually trip over or bump into something.

Predators prioritize stereoscopic vision over a broader visual field to precisely execute ambush attacks. Primates use it to navigate trees.[26]

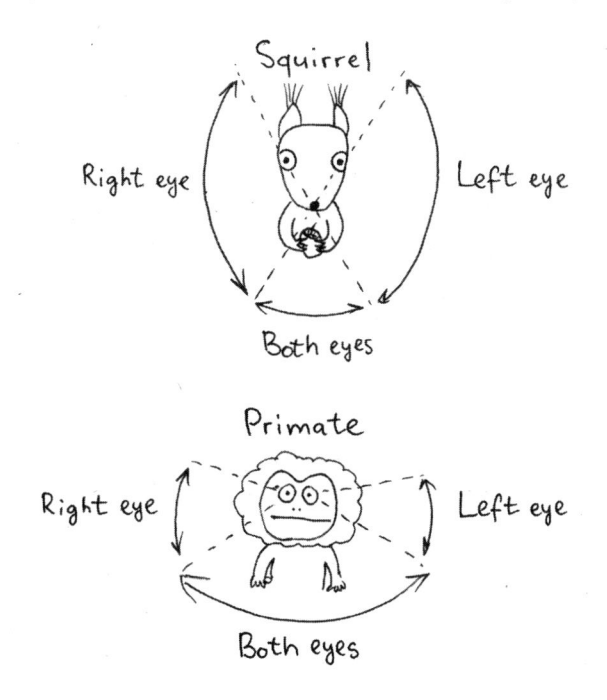

There was another consequence that arboreal life had on the primate body: the enlargement of arms, which in many branches of primates become progressively distinct from legs. Although sometimes primates use arms and legs interchangeably, they do tend to differentiate between them, especially when they are climbing vertically. In particular, apes, with their upright posture, prefer to use arms for climbing and legs for walking—chimps readily stand up on two legs and carry objects in their hands if what they are carrying is particularly valuable.[27] We are used to thinking of this feature as being unique to humans, but in fact, arms as specialized grabbing limbs precede our species by tens of millions of years.

But the most important change caused by a daytime, tree-dwelling lifestyle was in not in the primate body, but in their way of life itself.

The whole reason mammals had been nocturnal during the Mesozoic era was to avoid predators. Now in the Cenozoic, primates were back to daytime. But this also brought them back to the old problem: predators could now see them better, even if they were now different predators. And so virtually simultaneously with shifting to a daytime lifestyle, primates start living in groups.[28]

It is a profound notion that our sociality is built upon group defense. It isn't always like that—dolphins and orcas, for instance, form large groups primarily for the purpose of hunting. But in the case of primates, the deepest reason we seek out social bonds is to protect ourselves against the dangers of the outside world. Incredibly, this deep-seated evolutionary insecurity is once again seen in what oxytocin, that ancient "love hormone," does to us. There's a dark side to it. Oxytocin release in the brain increases trust and generosity, yes—but only, as it turns out, toward those who we perceive as our "in-group," our tribe.[29] At the same time, it actually increases distrust of others, the "out-group" (such as fans of a different soccer team).[30] We primates are neurologically wired for a tribal instinct.

But there's more. There's good reason to believe that group living, initially motivated by this need for protection, directly caused the evolution of human intelligence and, with it, every expression of culture, art, or science ever produced by our species.

What Made Us Human

In the past, many explanations of human uniqueness focused on *what gave us the ability* to become as intelligent as we are, rather than *why we would want* to be so intelligent. We often take it for granted that intelligence is what every animal obviously wants, and we just figured out a better evolutionary path toward it. One classic explanation for this involves, for example, walking on two legs, caused by a transition from trees to grasslands, which freed the hands from climbing and allowed us to do more complicated things. Another explanation focuses on our increasingly meat-based diet, which allowed for larger brain sizes. These factors certainly played critical roles in *allowing* us to become who we are. But they alone don't necessarily explain what is so good about being intelligent in the first place. We just assume that to be self-evident.

I think it's a bit of a self-serving assumption, like jellyfish wondering why no one else has managed to evolve stinging cells. We like to believe that we somehow won evolution—a notion we discussed in chapter 3 when talking about complexity and perfection. We have this image of an ape standing up, picking up a stick, and being rewarded for this achievement with a massive brain.

But the truth is, intelligence comes at a price, and for many species, the benefits just aren't worth it. A brain such as our own takes prodigious amounts of energy away from a body already burning through its fuel: a gram of brain tissue uses ten times the amount of nutrients as an average gram of the human body. Besides, a bigger brain is heavier and easier to damage. So there are considerable evolutionary costs to an enlarged brain. For any given species, these costs eventually outweigh the diminishing returns of brain enlargement. All brains have an evolutionary stage at which they are large enough. If a double-sized brain provided rhinos with a survival advantage, over millions of years their brain would have *certainly* doubled in size—you have to have very little awareness of evolutionary history to believe that we alone cracked some code that eluded everybody for eons. For rhinos, there wasn't any extra advantage in larger brains, so their brains turned out just as they did. The question is not why humans succeeded where others failed—as we tend to think—but why we needed supercomputers when others were fine with calculators.

There's an interesting pattern that may explain it. If you measure the size of the cerebral cortex—the brain's "machine of understanding"—in different primate species relative to the rest of their brain and plot it against the number of group members typical for each of those species, the two numbers fall on a straight line: the more members, the bigger the cortex. Humans are number one on both accounts—our cortex is the largest relative to the rest of the brain, as is our typical group size, estimated around 150—that's the number of people in a typical hunter-gatherer society and a typical cap on the number of active social acquaintances that we moderns can maintain. For example, corporate organizations often naturally fragment into units of about 150 people.[31]

Why would that be? This is far from a resolved question, but the proponents of the so-called social brain hypothesis say that reason is that social behavior is a uniquely demanding task, putting unprecedented strain on our brain's capacities. All mammals, to some extent, use their brain as a mirror, understanding others' behavior by modeling it inside their own mind. But primates, whose defensive groups swell into the tens and even hundreds, had to contend with tens and hundreds of these complex, interconnected models of other group members—their personalities, their emotions, their mutual relationships—which one of them did what to whom at what point and so on, a tremendous trove of complex data that we, humans, take to be as natural as eating dinner but that would befuddle even the smartest non-primate. In short, the social brain hypothesis states that *social life is what pushed us to become intelligent.*

The way this explanation differs from others is by offering an incentive rather than simply means to achieve it: yes, free hands, meat diet, and many other factors made our brain possible, but the reason we needed it in the first place was to remember all our friends who helped us fight monsters.

As cheesy as it sounds, I think about it all the time. There have been many different fables told about the birth of the human species: that it was work that made us human (this was the communist narrative—an ape picking up a tool) or maybe that it was violence (this is the narrative from *2001: A Space Odyssey*—an ape picking up a weapon). Those were not just scientific theories—they were origin stories, as important for a modern mind to make sense of itself as myths were to an ancient mind. An origin story is told to explain *what you are really about*, and in doing so, it doesn't simply describe the past but provides a template for the present. If you are *about work*, then work is the pillar on which your life should naturally stand. If you are *about violence*, then there is no sense trying to avoid it. But the more we learn about ourselves, the clearer it becomes that we are really *about others*. Our entire essence is to carry tens and even hundreds of peers inside our brains, to navigate the vicissitudes of their emotions and relationships, to derive both meaning and joy from living life together. It has long been recognized, for example, that happiness depends far less on individual well-being than on the richness of social contacts. Social life has a profound effect on us, and not just mentally but physically: for example, the Harvard Study of Adult Development, which began in 1938 and tracked hundreds of people for several decades, famously showed that close relationships are better predictors of long and happy lives than social class, IQ, or even genes. Too often, modern lives let us forget a firmly established fact: *friends are worth living for.* The social brain hypothesis puts an origin story behind this simple truth.

It also puts the birth of our species in a broader context. Our brains started swelling in size long before the first Homo sapiens. All primates share the relationship between group size and the cerebral cortex, which means that it always took a large brain to handle many peers.

And that, in turn, means that sooner or later, something like a human was inevitable.

When eukaryotes first started extracting energy from other organisms, this set the trajectory toward the human species—eventually there was bound to be someone who could control fire and even nuclear fission. There's something similar that the social brain hypothesis points to, at the deepest level. Once primates were swept in a drive to enlarge their groups and brains, eventually there was bound to be someone with groups large enough and with brains advanced enough to start talking to each other, inventing symbols and abstract categories—and from that, finally, there was bound to arise some form of culture, art, and civilization.

It is this final essence—an abstract, symbolic language passed from person to person by cultural transmission—that completes the design of a human being that we had seen gradually crystallize over billions of years. But to understand why language was so important for our species, we must now take a detour. Most books about human evolution begin right about here and proceed through the past few million years to the present, during which apes gradually evolved into several species of *Homo*, of which today only one survives—the "wise" one, or *sapiens*. But our quest instead takes us inward, into the human brain, into the sea of electrical signals pulsing through this astounding machine that runs our conscious minds.

Part III.
Where I Came From

Chapter 9. Animals of Abstraction

> Could baby cows be thinking?
> I saw the other day
> They had their tails swinging
> And stared so far away.
>
> AGNIA BARTO

At any given second—even when we are asleep—millions of neurons in our brain erupt in a cannonade of electrical explosions, or spikes, which carry signals from neuron to neuron, ricocheting through entangled networks of synapses like lightning-fast electrical pinballs, their staccato rhythms coalescing into an ebbing and flowing cloud of information. With the right equipment, you can plug a wire into a neuron and record these explosions, the silent drumbeat of cognition. You can actually convert the electrical pulses into sound and listen to them—it's a convenient way to keep track of what's going on during an experiment. The most commonly recorded neurons, mouse neurons, make a rather unpleasant popping noise, but our lab's specialty is not mice but sea slugs, and their spikes sound like a satisfying thud of a kick drum. You can use the connected wire to direct electrical charge into those neurons, and in response they produce a steady beat, which really does sound like house music.

This thud, the neuronal spike, is one of the most elemental, stripped-down manifestations of cognition—the process that we generally refer to as *thinking*. We don't tend to put philosophical weight into cells and molecules, but a spike is a perfect place to do just that. It represents an essential bridge between a neuron and a mind—more broadly, between the physics and chemistry of living matter and our inner lives. It is in search of such bridges between small and big, between mundane and profound, between specific and abstract, that our lab studies mollusks.

To explain this, I have to backtrack a bit. In fact, I have to backtrack quite a lot—more than half a billion years into the past from the last chapter, where we left our ancestors, the social primates that were just about to invent language and finally become humans. It is at that distant point in time—just before the Cambrian explosion—that the last common ancestor of humans and sea slugs roamed the bottom of the seafloor. Actually, almost all types of modern animals, with the exception of sponges, cnidarians, and a few exotic offshoots, are descended from this ancestor. It was probably a small worm, which may have been a bottom dweller or a burrower.

Anything that humans and sea slugs share, that ancestor must have had, too.* Anything that is different between us, we acquired later, after the human and slug branches of evolution split from each other. So in a sense, to study a sea slug from the point of view of a human is to study that distant, Precambrian worm that was the prototype of all subsequent animals. To find common ground between slugs and humans, two species so utterly distinct in physiology and cognition, is to distill the essence of what physiology and cognition are—to strip down the layers, to reveal the original design. Working with sea slugs makes you think in philosophical generalities rather than technical specifics.

For example—and I am sorry about the morbid details—a slug can be cut up into pieces, and the pieces will still work together if you leave intact the nerves between them. This is known, with a dash of dark humor, as a "semi-intact preparation."[1] When exactly does a semi-intact slug become not intact

---

*There is also a possibility of convergent evolution—some ideas can be independently developed by two separate branches of evolution. For example, both humans and insects have independently evolved legs, so although we share this trait, we haven't acquired it from a common ancestor. But this is an exception rather than the rule, and usually (although not always) the two situations are easy to tell apart.

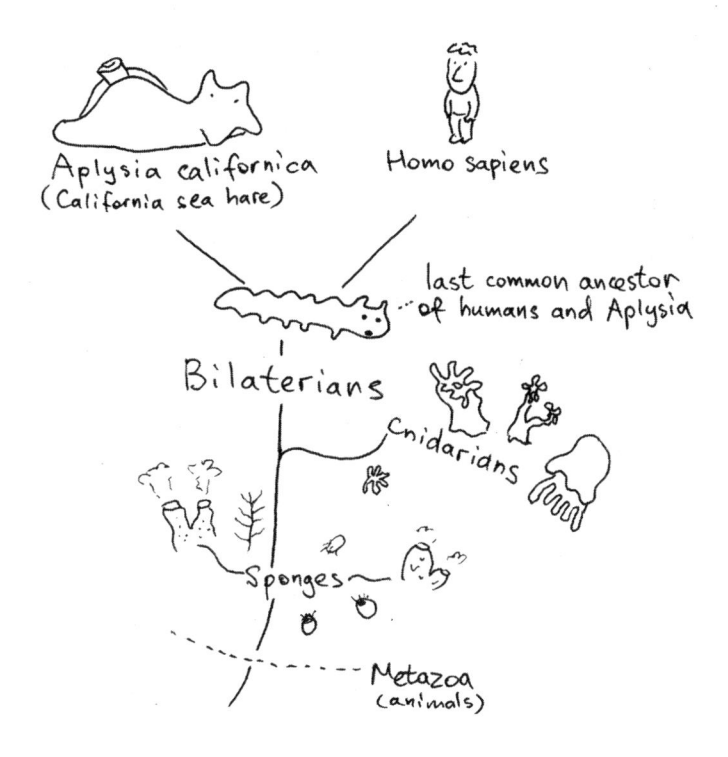

at all—as in, when does it die? There isn't a real answer. As you deconstruct the animal, you gradually lose some functionality—the semi-intact prep can no longer move around and forage for food, and if you further separate it into cells, it won't have any reflexes because there will be no body parts left—but put neurons in a petri dish, and they continue business as usual. Where you draw the line is really up to you. It turns out that what we think of as the sharpest distinction known to nature—alive versus dead—is a lot fuzzier than it seems, and you run into this philosophical conundrum on day one of working with *Aplysia*—that's what our sea slugs are called.

What studying distant species can teach us about the human experience is *what it really means*—not just from our self-centered perspective, but from much broader vantage points. Sea slugs are a way to escape the rabbit hole of complexity that is the human species and instead focus on the purest essences of animal life and cognition.

So then, what does it mean *to think*?

Preserving Patterns

The most important and most difficult task of the brain is to see through the chaos of immediate experience, which is messy and unpredictable. To be efficient and safe, an animal must be guided by more stable trends and a broader outlook than what the environment offers at any given moment. Brains take in this chaotic world, find within it *abstract ideas*—patterns in the chaos—and use them to decide what to do next.

Take, again, the sea slug. If you gently poke the slug in one of its body parts (we actually use a water jet from a dental flossing apparatus), the animal pulls in that part—say, its tail or head. But in either case, it also pulls in the siphon—a breathing organ on the slug's back, which is particularly important to protect. How this happens is well known. There are a few hundred sensory neurons in the brain of the slug,* each corresponding to a particular location of the body. They have long arms (dendrites) that stretch from the brain to various locations on the slug's skin. When you apply pressure to the skin, a particular neuron senses it through its dendrite and becomes activated: a tail sensory neuron if you poke the tail, a head sensory neuron if you poke the head. These neurons, in turn, activate the nearby motor neurons, which are also dedicated to specific body parts and send signals back to muscles in the tail or head, causing them to contract. But at the same time, both sensory neurons send an additional signal to another motor neuron: the siphon-pulling one. So the tail and the head contract independently of each other, but the siphon contracts in either case.[2]

Each neuron is activated for a particular real-life reason. So you could say that their activities have different meanings. The sensory and motor neurons of the tail and the head represent, in their activity, the meanings *touch to the tail* and *touch to the head*. But the neuron pulling the siphon gets activated in both those situations and more: touch to the tail or the head, touch to the side, touch to the siphon itself, and so on. So its activity means something more general: *touch to the body regardless of location*.

---

*Many a nature show would tell you that sea slugs don't have brains—they have *ganglia*, but it's really just a different word for the same thing. Ganglia are like a few minibrains that are slightly more separated from each other in space than the parts of our brain. (In Russian, on the other hand, the spinal cord is called "spinal brain," so it's all a matter of convention.) I keep it simple and use the terms *brain* and *nervous system* interchangeably.

touch to the tail

siphon (Breathing organ)

muscle contractions

touch to the head

Defensive withdrawal:

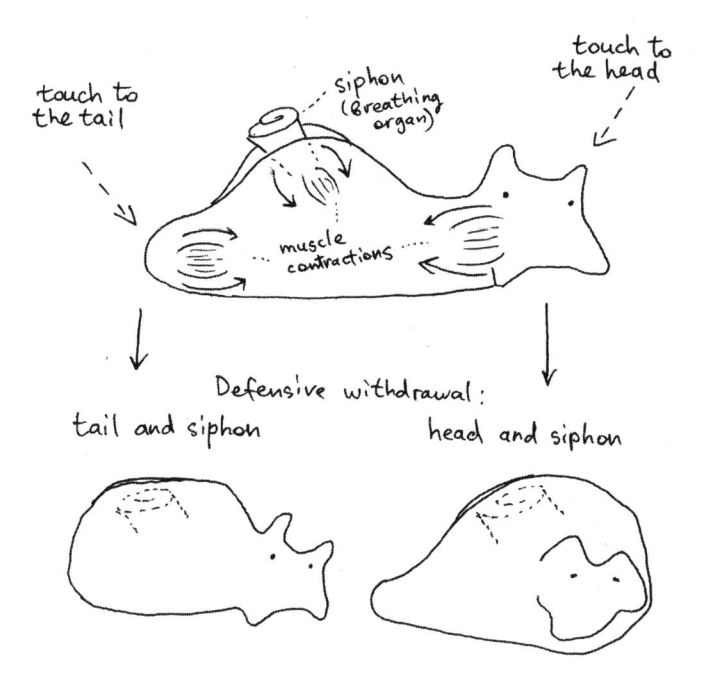

tail and siphon

head and siphon

What's more, the specific meanings that create this generalized meaning are not simply averaged, but mixed in various proportions, depending on the slug's past history. If you repeatedly harass the slug's tail using an electric shock, the siphon becomes more responsive to a tail touch, but not to a head touch—the animal learns where the danger usually comes from. So the abstract idea represented by the siphon-pulling neuron is not just *any* touch regardless of location,

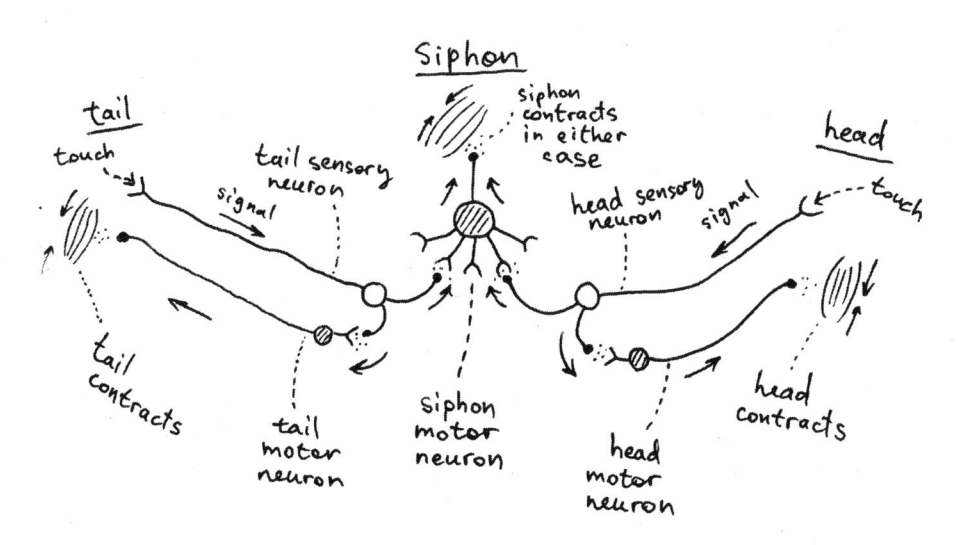

tail touch — tail sensory neuron signal — Siphon: siphon contracts in either case — head sensory neuron signal — head touch

tail contracts — tail motor neuron — siphon motor neuron — head motor neuron — head contracts

Animals of Abstraction

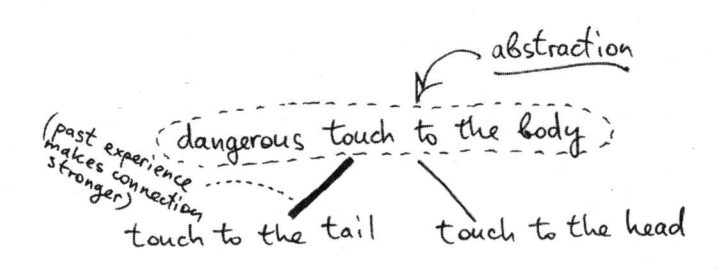

[handwritten diagram: "abstraction" with arrow pointing to "dangerous touch to the body", which branches to "touch to the tail" and "touch to the head"; annotation "(past experience makes connections stronger)"]

but specifically *dangerous* touch regardless of location. It's pure abstraction: there isn't *physically* such a thing as touch without location, and our water flossing jet is not *physically* damaging, regardless of where it is directed. But it's useful for a slug to consider this imaginary generalization based on past history and make it the trigger for siphon withdrawal, because the siphon should be withdrawn any time there's a possibility of a threat. So the slug uses this abstraction of dangerous touch, biased toward the most threatened body parts, to consistently make good choices in a large variety of specific situations.

Compare this to how our minds work. As a Russian, I am very fond of foraging mushrooms. Non-Russian friends who occasionally join me are horrified to realize that I don't actually carry a catalog of edible and inedible mushrooms in my head. (It adds spice to dinner.) What I have in my head is an abstract idea of an edible mushroom and an abstract idea of an inedible mushroom, with smaller, less abstract ideas plugged into them: gills can be OK but you need to be careful; spongy undersides are safer but watch out for strange colors. No two mushrooms are exactly alike, so I approach the process intuitively and holistically: I just look at a mushroom, and it either looks good, or it doesn't. Most mushrooms I don't even consider, but the ones that give me a good overlap of familiar traits, I pick.

How did I develop this fungal sixth sense? I spent some formative summers walking around forests with my grandparents, being shown good mushrooms and bad mushrooms. Over time, I memorized consistent patterns. Specific ideas about the mushrooms, like colors and shapes, became associated with more abstract ideas, like their edibility. It is probably one of the few bits of purely oral tradition that I carry—I can't imagine that any of my ancestors had ever used textbooks to ID their chanterelles.

So sea slugs and humans both use abstractions to guide our choices. In fact, our entire mental process can be visualized as a back-and-forth motion

abstractions

edible mushroom   inedible mushroom

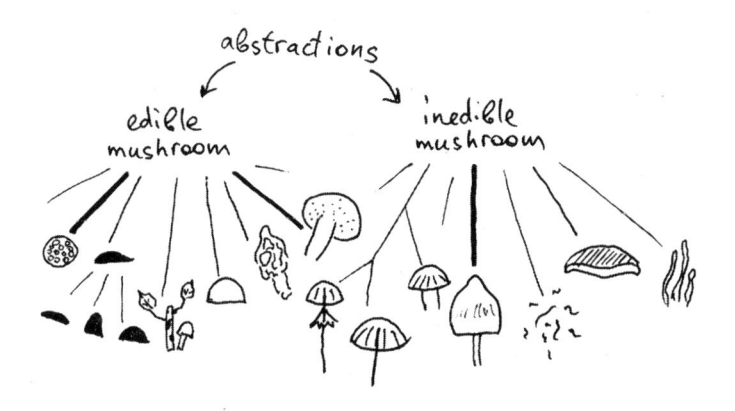

between specifics and abstractions of various levels. Everything in our minds is organized into a nested system of categories, like folders containing folders, containing other folders, and so on almost to infinity. Although a physicist would tell you that everything in the world is made of the same energy, atoms, and forces as everything else, in our minds we see it all broken down into specific objects, concepts, processes. We see distinct humans with distinct faces and distinct facial expressions. We see distinct animals split into different species. We see streets divided into roads and sidewalks, making our dogs puzzle over our strange walking patterns. Even the spectrum of light—a quintessential continuum without borders—is subdivided in our mind into six or seven colors, and no matter how hard we try, we cannot unsee the distinctions. Each of our mental categories is further subdivided, usually in many overlapping ways—humans, for example, into friends and enemies, old and young, fashionable and unfashionable, bus drivers, nurses, scientists, and tailors, American, Russian, and Chinese. Conversely, each category is part of a broader one: for example, I am convinced that deep down we all have a mental folder where we put both cars and horses, something like *big benign thing you can ride but can't get in its way*—otherwise, I find it too much of a coincidence that all cars have two headlights, positioned like eyes, and four wheels, positioned like legs—cars would look weird otherwise, wouldn't they?* Our mental space is like a vast, profusely branching tree, except it grows upside down, from the branches to

---

*It is particularly amusing that the first automobiles all looked like horseless carriages—that was the accepted abstraction of the time, *big boxy thing with wheels*—and then, over two hundred years, gradually streamlined design and increasingly organic curves made them look a lot more like carriageless horses.

the trunk, as we find patterns in the world, unify specifics into abstractions, and those into more abstract abstractions, and so on until we find the most abstract connections among all our experiences that hold them all together.

Whatever is going on through your mind can be thought of as motion through this tree, whether you are moving through familiar branches or creating new ones. When you have an idea—I just thought of a way to fix the leaky toilet with a chopstick!—you take two lower-order ideas—leaky toilet and chopstick—and find a pattern of their interaction that can solve your problem. You create a new folder. When something triggers an association—for example, the smell of borsch transports you from a Russian restaurant to babushka's kitchen—your brain suddenly finds that a common folder unifying the two environments already exists. When you come home from work and perceive many changes around you—time of day, location, people, furniture, lighting—your mind hops from the folder "work" to a different folder "home" and so you kick off your shoes and relax. We all have at some point been confused about these unspoken mental folders: for example, if you have colleagues over at your place, you might find it hard to convince yourself that the folder is "home" rather than "work" and find it hard to relax even if your colleagues are perfectly fun.

One interesting consequence of our mind being organized into this upside-down tree is that there's a difference between climbing upward and downward. For example, any elementary school teacher would tell you that subtraction is harder than addition.[3] Try to slow down your thoughts and pay attention to what, precisely, happens when you calculate this: 15 − 9 = ?

If you are like me, you look at 15 and at 9 and then start looking for a number to replace the question mark. What you *actually remember*, you might realize, is the sums: 9 + 4 = 13; 9 + 5 = 14; and so on. So you start trying these different numbers out: does this sum make 15? No. Does this one? No—until you find the correct answer, 6. This is how children first learn to subtract.[4] When I do it as an adult, it happens very fast, but I still notice a slight speedbump—it requires a lot more mental energy than addition, which proceeds in one click.

This is because addition works from small to big—from specific to general—whereas subtraction works from big to small, general to specific, and that, generally speaking, is a harder direction for the human mind to travel.

Addition   Subtraction

(?)   (15)
↗ ↖   ↙ ↘
(9) (6)   (9) (?)

Abstraction, at very different levels of complexity, is the core principle that unifies our cognition with the cognition of a sea slug. Such a deep unity means deep evolutionary roots—that Precambrian worm, which gave rise to almost all modern animals, must have been capable of abstraction too. This ability is not specific to any one nervous system or any one animal. To truly reduce the essence of cognition to its most basic elements, we must dive deeper into the common denominator of all nervous systems: the neuron.

Decision Time

Abstraction is what a neuron does for a living—you can see it in the very design of the cell. Neurons are cells that have a central body and long, branching arms called dendrites and axons. Dendrites receive signals from other neurons, axons send them out to other neurons. So each neuron has inputs and outputs. Neurons constantly evaluate the pattern of inputs and decide whether they should trigger the outputs. This very act is abstraction—as pure as it gets.[5]

This is how it works. Neurons communicate with one another using chemicals called neurotransmitters, which are released from an output ending of one neuron and sensed by the input ending of another—these junctures are called synapses. The most garden-variety neurotransmitter, glutamate, causes the opening of ion channels on the membrane of the receiving neuron, which momentarily floods it with positive charge, like a small injection of electricity. But this only happens in one synapse at a time, and for the receiving cell that could have a thousand such input synapses, this one pulse of positive charge usually means little—it is quickly cleared away. However, if many of a neuron's input endings are simultaneously getting these small electrical injections or if some of them are getting the injections at a very high rate, the small effects start to add up. The positive charge begins to creep up through the dendrite into the body of the neuron and, when the electrical field brought by this charge

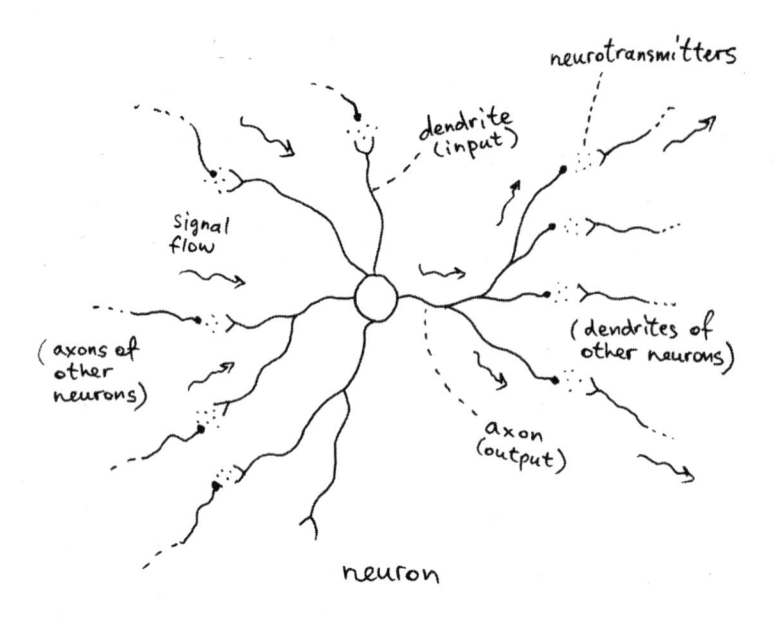

[Diagram of a neuron with labels: neurotransmitters, dendrite (input), signal flow, (axons of other neurons), (dendrites of other neurons), axon (output), neuron]

finally reaches a threshold, it triggers an all-or-none burst of electricity—the spike, also known as an action potential. Once triggered, the spike rips the membrane of the cell open: the same electricity that was slowly oozing through the synapses now bursts through, like a millisecond-long tsunami, and just as suddenly, it hits a wall and abruptly recedes. This is what sounds like a thud when you convert it into sound. It is an electrical pulse that almost instantly reaches every output ending of the triggered neuron and releases neurotransmitters from all of them simultaneously. This passes the signal to the next rank of neurons, which repeats the entire process.

The proper scientific term for a spike, *action potential*, captures well the essence. A neuron only has one *action*: this burst of electricity, which spreads rapidly within it (*potential* refers to the electrical field traveling along the neuronal membrane) and releases neurotransmitters as a final crescendo. The neuron either *acts*, or it doesn't. The spikes themselves never get weaker or stronger. The only thing that changes is the rhythm at which neurons fire them. When we plug wires into slug neurons, we can artificially direct a stream of electricity into them, imitating synaptic input. There's a knob for that on our machine, and as you dial it up or down, the kick drum in the speaker starts beating faster or slower but staying the same volume and timbre. (I confess there have been a few times I could not resist the temptation to beatmatch the neuron with a

neurotransmitter released — receptor detects neurotransmitter — signal flow — opens ion channel — ion channel (closed) (open) — positive charge enters the cell

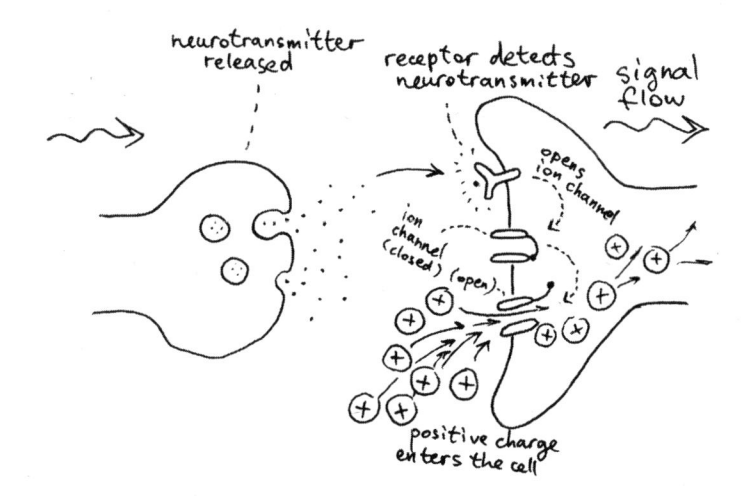

song playing on the radio.) It's just like that in real life—neurons converting the electricity they receive into their own rhythm—except the electrical charge coming from actual synaptic inputs is not a steady stream but a subtle, often barely noticeable trickle, seeping in through tiny ion channels unlocked one at a time by neurotransmitters of other neurons, themselves pulsating with unpredictable activity.

In this way, a typical neuron collects and evaluates thousands of signals, constantly deciding whether or not they are sufficient to produce a signal of its own. So, when that does finally happen, this signal, by its own nature, represents a more abstract piece of information than the preceding signals that triggered it. It then joins many other signals trying collectively to trigger a spike in the next neuron. So, as information moves deeper into a brain, at each synaptic juncture it becomes more and more abstract. A neuron is a cell that is built to abstract, and a spike is embodied abstraction.

Digital computers also abstract information. Today's CPUs and GPUs are built using semiconductor technology—the *silicon* in *Silicon Valley*. A basic building block of a silicon microchip is a transistor—a sort of digital neuron. Like a neuron, it takes in multiple inputs—albeit only two in this case. Electrical charge from input 1 goes straight to the output, but on its way, it must pass through a piece of silicon connected

input 1 →
input 2 →
input 1 → input 2 → output

silicon transistor

Animals of Abstraction **171**

below threshold

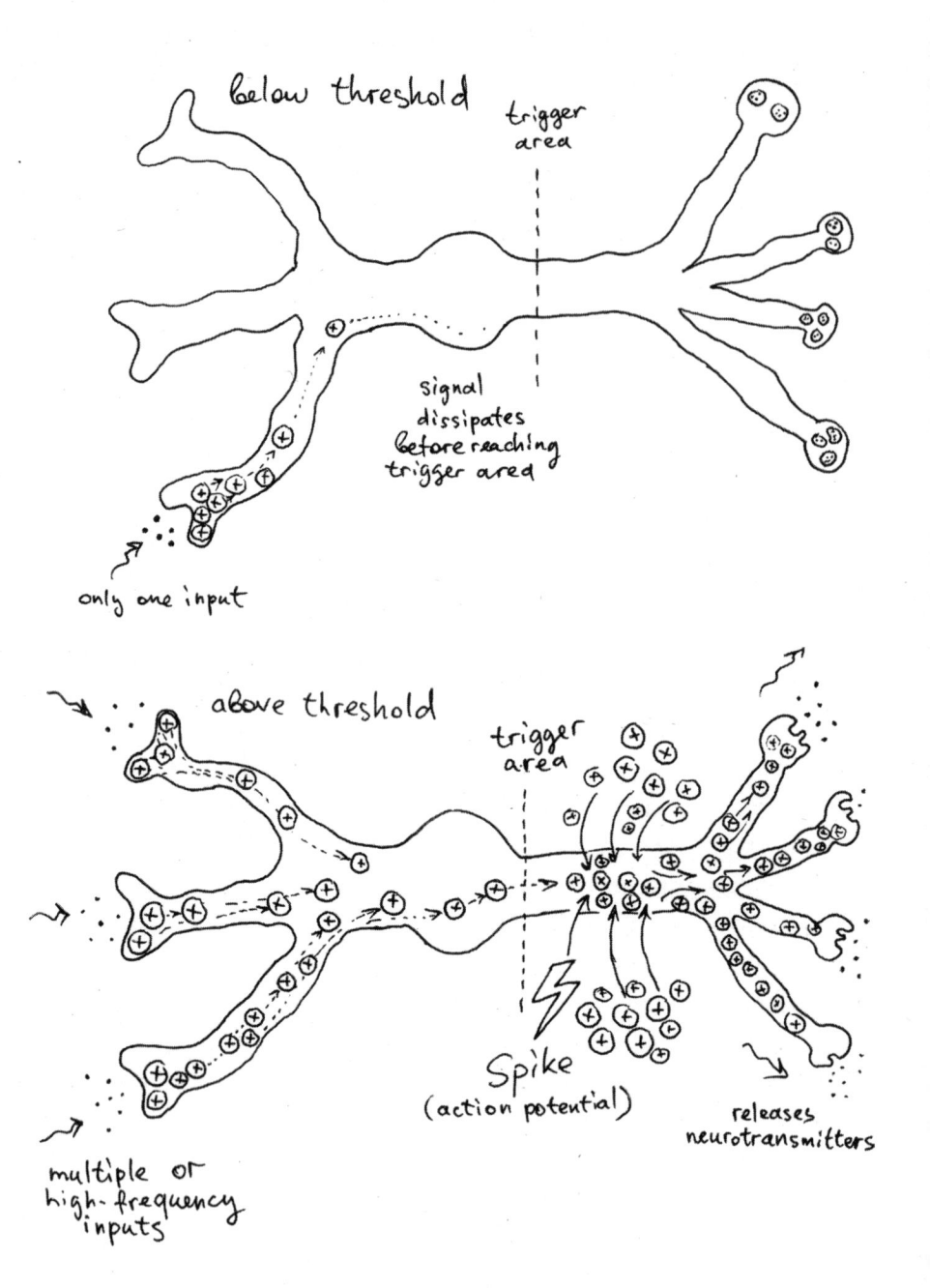

trigger area

signal dissipates before reaching trigger area

only one input

above threshold

trigger area

Spike (action potential)

releases neurotransmitters

multiple or high-frequency inputs

to input 2. When input 2 is on, silicon is a conductor, and when input 2 is off, it's not (that's why it's called a semiconductor), and input 1 can't pass. So both inputs must be on to have any output. That's a pattern, and if it is detected, an output is produced. You can then plug the outputs of two separate transistors into a next-level transistor, which would detect a pattern of patterns, and so on. There are slightly different transistors that detect slightly different patterns, and silicon microchips contain billions of them all plugged into each other in various combinations.

But aside from this common ground in abstraction, there are a few big differences between digital computers and brains. One of the key distinctions is the way they handle memory. For a computer, information processing is separated from information storage. A CPU always stays the same—any new information is recorded in a separate medium, such as the computer's hard drive. A computer takes in a large amount of information, performs on it the same, predetermined series of abstractions, and records the result in the end. In a brain, however, information is processed and stored by the same physical medium—the neuronal network. Our cognition is inseparable from memory. They happen not in stages, but simultaneously, seamlessly. This allows us to not only process information in the moment, but dwell on it. Signals pass through neurons on the scales of milliseconds, but our thoughts last much longer. They linger, fester, and build upon each other, constantly updating the brain, channeling the ongoing present through the memories of the past in order to imagine the future. This is part of the reason humans and computers excel at different tasks—silicon microchips are much better at doing standardized tasks, whereas our brains are much better at flexible, nuanced, context-dependent ones. In a brain, abstraction and memory are part of the same process. As we abstract reality, again and again, we gradually memorize the patterns of abstraction, which become our lanes of thinking.

It is one of my favorite facts in the history of science that the molecular mechanisms behind both of these fundamental brain abilities—abstraction and memory—were first described in mollusks. Who would have thought? So far, we focused on abstraction, embodied in physical form by action potentials. These were famously described by Alan Hodgkin and Andrew Huxley in the giant axon of the squid—for this seminal work they received a Nobel Prize. Memory, the essential counterpart to abstraction in our minds, is embodied in another neuronal process: synaptic plasticity. This process, too, was first studied in mollusks—our friends, *Aplysia* sea slugs. *Aplysia* was pioneered as a research model by Eric Kandel, who also received a Nobel Prize for uncovering the most basic mechanisms behind synaptic plasticity and who was a mentor to my mentor, Tom Carew. To this day, the main reason why we study sea slugs is their memory.

Do You Remember Anything?

The very notion of slug memory surprises a lot of people. There's a meme regularly sent to me based on a real study in which scientists erased certain memories from a snail's brain: in the meme, scientists ask the snail if it remembers anything, it does not respond, and they go, "my god what have we done."

Do slugs even have memory? You have to explain to the naysayers that what "memory" means for a sea slug is different from the memory we have in mind when we close our eyes and think about what we did last week. Sea slug memories are much simpler: say, you give it a few electric jolts to the tail, and that makes it jumpy for a day or two; you touch it, and the siphon pulls in more vigorously than before. Sure, say most people, but that's not *really* memory. It's reflexes, adaptation, plasticity—but it's pointless to give this siphon movement the same name that we give to our rich, dreamlike internal recall.

But there is, in fact, a point in using the same word, *memory*, for those very different reactions. It's not just a matter of drawing parallels, like when you say, "leaves are nature's solar panels." Leaves and solar panels don't *really* have a common origin or work in the same way, so this parallel is metaphorical. Memory is different. When I say that curling into a ball after yesterday's shocks is the sea slug's equivalent of a human thinking back to what they did last week, I mean that literally. Of course there are lots of distinctions—all those layers that have been added on both our nervous systems since humans and slugs split from each other in our lines of descent hundreds of millions of years ago. But deep inside those layers, the same gears and cogs are turning. If you shrunk yourself to the size of a molecule and traveled through the neuron of a sea slug, a human, or a mouse, you wouldn't even immediately know that there was any difference. Yes, eventually you would figure out that slug neurons are larger, slower, and have fewer contacts than human neurons, but you would have no doubt that these complex machines are doing fundamentally the same things using the same components.

You can actually simplify beyond a sea slug and still find memory—just as we did earlier, when we found abstraction in the most basic behavior of a neuron, a spike. Remember how slugs pull in their siphon when you touch either the head or the tail? The head and tail sensory neurons connect to the same motor neuron that pulls the siphon. You can extract all three of those neurons from the slug's brain, place them in a petri dish next to each other, and a few days later they will form two separate synapses: head-siphon and tail-siphon, just as they used to in the animal—as if nothing happened. If you then take a small pipette and puff neurotransmitter onto just one of the synapses, it will grow in strength compared to the other one, and as a result, the tail sensory neuron will then start triggering the siphon neuron more vigorously. Depending on how many times you do the puff—which emulates electric shock to the animal—the boost to the synapse persists for anywhere between a few minutes to a few days.[6]

This is what synaptic plasticity is: the ability of synapses to change, independently of each other, in response to experience, to incorporate past history into their physical configuration. It is the canvas on which all our memories are painted.

Most of my colleagues would feel more comfortable saying that this process *underlies* memory or *serves as a mechanism* for memory—which, some of them might say, is a concept that can apply only to a behaving animal. Personally, I have no reservations saying that synaptic plasticity *is* memory, just as it is in complete slugs, in mice, or in humans. Seeing how a sea slug can be deconstructed and reconstructed into components of various degrees of intactness, it's hard for me to justify the notion that memory only appears once the animal is fully assembled. Molecules that disrupt memory also disrupt synaptic plasticity—for example, protein synthesis inhibitors selectively block long-term but not short-term memory in mice[7] and sea slugs,[8] and in the petri dish with neurons, they block the long-term boost of the synapse but not the short-term one.[9] Other molecules—for example, growth factors—have the opposite, enhancing effect in all cases.[10] A mouse memorizing a maze, a slug memorizing a shock to the tail, and a neuron memorizing a puff of neurotransmitter use the same molecular tools derived from the same evolutionary source for the same essential purpose. It's just a question of how many neurons are put together and whether or not you also connect them to eyes, ears, muscles, tails, and siphons. At the root of the process is neurons and synapses changing in response to their past history.

There are many different patterns of experience that can cause a synapse to grow or shrink, but the simplest and most common way to give a synapse a prolonged boost is the one I already mentioned—repetition. Whatever the synapse is used for and wherever in the brain it is positioned, as a rule, if it is used a lot, it grows—both at the receiving and the sending ends. The electrical "punch" that it delivers from the sending neuron to the receiving one increases. The synapse becomes more influential among others. Conversely, if the synapse is not used, it atrophies, like a muscle—the same "use it or lose it" principle applies in both cases.

What is especially interesting about this detection of repetition by synapses is that they can distinguish *again* from *more*—almost universally, a stronger memory forms if you learn something *repeatedly*, with rest periods, compared to learning it for a longer time all at once.[11] This makes sense for a wild animal: events that repeat consistently over a long period of time are usually worth remembering better than one-off flukes, even if the singular event is more intense. Some neurons, as our research shows, can even distinguish escalating repetition from waning repetition and strengthen their synapses only in response to the former.[12] This also makes sense: if a threat or a reward are increasing in intensity over time, that is worth remembering more than a threat and reward that are gradually disappearing.

In other words, learning and memory formation, even by individual neurons, are also a form of pattern finding, just like abstraction. Except abstraction is finding patterns among ongoing things, whereas learning is finding patterns in time.

Thinking is using both forms of pattern finding at the same time—nothing less, nothing more.

Inward and Onward

A sharp border separating the physical and the mental is intuitively obvious. It is the reason why most cultures believe in a soul of some kind—an entity distinct from the body that experiences its sensations and beliefs.

Science, too, has generally seen this border as impenetrable. The great German mathematician Gottfried Leibniz used the famous "mill argument":

if there was, he wrote in 1714, "a machine whose structure makes it think, sense, and have perceptions," then you could imagine enlarging it to a size of a mill and walking into it—but all you'd be able to see is parts that push one another, and no amount of detail could explain how all of that converts into actual thinking and perceiving.

In the mid-twentieth century, this wisdom was considered so unshakeable that an entire tradition of experimental psychology was premised on basically admitting defeat: behaviorism. It stated that behavior—that is, the physical movements performed by an animal body—is the only readout for an animal mind there could possibly be, and "mind," as a scientific concept, should never even be discussed: the goal is merely to determine which inputs into the "black box" of the brain predict which outputs come out of it—what stimuli cause what behavior. The rest was pseudoscience.

But then, as time went on, neuroscientists—as experimental psychologists came to be known in the twenty-first century—*did* get into the black box. Just as microscopy once illuminated previously inaccessible worlds, brain imaging technologies opened the window into the inner workings of the brain—not just its behavioral output. Brains, it became clear, are not just relay stations that route sensory stimuli to muscles—they also *create meanings*. By abstracting and memorizing complex patterns over long periods of time, neurons can represent in their activity ideas that do not have a specific physical counterpart in the outside world. Today, using tools such as fMRI, we can often pinpoint extremely specific meanings of these abstract ideas—individual words, personalities, concepts—to small clusters of neurons or, in research animals, even to individual brain cells. We now have technologies to create false memories, to turn them on or off at will, to see them take hold in an animal's brain without the animal even having to do anything. We can decipher the *content* of what goes through an animal's mind. Maybe the most quintessential case is place cells—neurons that are associated with a specific location and that fire when an animal goes there or even thinks about it. For example, when a rat sleeps, its brain replays the sequence of place cells that track its past movements—it's a dream you can decipher without any ongoing behavior.[13] This gradual movement of brain science away from the dogma of behaviorism—that behavior is the only thing that's truly real—is sometimes termed the "cognitive revolution."

The wall between the physical and the mental remains, however, even though it has become palpably thinner.

I often use the broadly descriptive word "mind" to encompass the processes that Leibniz called "think, sense, and have perceptions." In today's science, this includes two distinct concepts: cognition and consciousness. Cognition is information processing: memory and abstraction. Today, no one would have a problem saying that a machine has cognition.

Consciousness, however, is something else: the first-personness of it all, the fact that it all *feels like something*. Consciousness, for most people, remains on the other side of the wall. The wall is vigilantly guarded by its staunchest defender, the philosopher David Chalmers. Chalmers is famous for introducing "the hard problem of consciousness," which is, roughly speaking, a modern version of Leibniz's mill argument.[14] No amount of knowledge about the objective nature of the brain, says Chalmers, can explain the mysterious nature of subjective experience.

My own view on consciousness most closely aligns with that of cognitive neuroscientists Karl Friston and Andy Clark, whose theory of hierarchical predictive processing is discussed in more detail in the next chapter. Friston and Clark stand opposed to Chalmers: they argue the wall between the physical and the mental, the objective and the subjective, doesn't exist. The brain is a biological machine with built-in abstraction and memory that runs sophisticated software. This software is constantly trying to understand ongoing experience, and consciousness naturally arises as part of this process of understanding—more on this soon. We are yet to understand fully how all of the parts work—but once we do, say Friston and Clark, there's no additional mystery, no impenetrable wall beyond which we can't reach.

The real problem, which they call "the meta problem of consciousness," is to explain the *idea* we have that there's something else to our consciousness—the conviction that *first-personness* is distinct from simply *being*. What needs to be explained is not the hard problem of consciousness, but why David Chalmers believes it is a problem.[15]

To say this debate is unresolved is an understatement.

I agree with Friston and Clark that "first-personness" evaporates if you meditate on it for long enough. There is no need to explain how the objective thing transitions into the subjective thing, because there really aren't two things. There is no wall.

What I find curious is that Friston and Clark came to this conclusion by studying the human mind, whereas I arrived at the same place by studying sea slug neurons. You could say we approached the wall from both directions and met in the middle without ever finding it.

When I described how sea slugs abstract information—how the neuronal signals that mean *touch to the tail* and *touch to the head* combine into a more general meaning, *dangerous touch to the body*—I deliberately avoided labeling these meanings as "ideas" or "essences." *Essence* is the word I have been using throughout this book to describe "ideas of nature"—the products of evolution's ingenuity, embodied the world itself. *Idea*, on the other hand, implies meaning that exists in our imagination, a logical product of our brain activity. *Essences* are objective, *ideas* are subjective. So where do abstractions in the slug's brain belong? Are they essences or ideas?

From our human perspective, external to the slug, they are essences—just like the essence of a leg is to walk, the essence of a neuronal signal can be *dangerous touch to the body*.

But if we had a way to look at the world from the perspective of the slug, we would have no reason *not* to call these essences ideas. From this internal sea slug point of view, we would know nothing of the neurons that move our siphons—we would just know of the potential danger in the environment when someone touches our tail.

Sure, you might say, but that's not *really* an idea. It's an "idea," in quotation marks, at best—a metaphorical description of how a sea slug responds to a stimulus, and it is pointless to give our rich, subjective mental constructions the same name.

But this is exactly how most people think about sea slug memory, too—it's "memory" in quotes, not real memory. And they are wrong—our memory is, in fact, the same process, just taken to an extreme of complexity.

What if our intuition here is just as wrong as with memory? It equally doesn't make sense that sea slug memory, human memory, and the memory of neurons in a petri dish share something in common, and yet they do. What if this is the same? What if the subjective is just a complicated form of the objective? *What if all ideas really are essences?*

If this is so, then the wall between the physical and mental is not really a wall at all, but simply a matter of perspective—hardware and software. You can think of neuronal spikes and synaptic plasticity as properties of either, in the same way that you can think of a phone app either as lines of code or as electrons coursing through a microchip—both are correct, at the same time. And depending on which way you think about spikes and synaptic plasticity, they are either external or internal to you.

What studying sea slugs has taught me is that our understanding of human experience is often clouded by the extraordinary complexity of our brains.

Once you disregard the complexity, many things that puzzle us about it—like the nature of life and death—simply dissolve, like a mirage. And one of those puzzling things that dissolves when you look at it closely is the distinction between the objective and the subjective.

It is a very difficult mirage to unsee, just as it is difficult to not see the rainbow as seven separate colors. But in the remaining chapters of the book, we will do our best to let go of it.

Chapter 10.
Fire from Within

Why do you pretend to be
Now the wind, now a rock, now a bird?
ANNA AKHMATOVA

Comparing brains to computers is today's equivalent of comparing humans to apes in Darwin's era—it upsets a lot of people.

In a way, though, we *like* to believe that our brains operate like very advanced computers: you input some information; your brain analyzes it and comes to a decision. It is a comfortable belief because it gives us confidence in our own judgment. All we have to do is get the calculations straight, and we'll be perceiving the world for what it *really is*.

But consider a visual illusion like the one known as "My Wife and My Mother-in-Law."

It might take a few seconds, but you quickly figure out there are two ways to see this image.

Either you see a younger woman turned away, or you see a profile of an older woman looking to the left.

There's nothing mysterious about it—you can look closely at the image and see that the jawline of the "wife" is simultaneously the nose of the "mother-in-law," that one woman's ear is another one's eye, and so on. It's a clever visual trick, but that's that. Once you figure it out, you have all the information.

And yet, no matter how hard you try, *you can't see both women at the same time*. If you really focus on that task, you can sense your brain kicking into high gear, rapidly shifting between the two interpretations so that they almost fuse together like pictures in a flipbook making a cartoon. But they never truly combine. The brain simply refuses to see this image for what it really is—lines on paper rather than one woman or another.

This seems like a big difference between humans and computers. Machines are supposed to make cold calculations, to take themselves out of the equation and consider only the facts they are given. We, on the other hand, appear to actively participate in our own perception. *To see, we need to decide what we want to see.*

In the previous chapter, we saw that abstraction was the main operating principle of both computers and human brains. In general, as information flows through a network of neurons (or transistors), it becomes more and more abstract. We also saw that there's a big distinction in how our brains and computers handle memory.

Here's another distinction. In a simple digital computer, such as a pocket calculator, the flow of information goes in only one direction: from input, to processor, to output. In the human brain, information flows in two directions at the same time.

For example, a calculator takes in small, granular pieces of information—individual digits like "1" or "5," "2" or "9," which you type in. It converts them into bigger, more abstract pieces of information that you can't input directly—numbers like *15* or *29*. It then takes those virtual entities and derives an even more virtual entity out of them, like their product, 435. That's where the calculation stops. The calculator sends this result out into the world to become real—"435" on your screen.

This is also how sea slugs operate, if you recall the previous chapter. They receive an input (touch to the tail or head), derive a more abstract entity out of it (dangerous touch to the body), and then connect that abstract entity to an output: pulling in the breathing organ.

We, too, receive inputs—a never-ending stream of signals from the outside world delivered into the brain through the eyes, the ears, the taste buds, the nerve endings in the skin, the chemical receptors in the nose. We, too, extract from them abstract ideas, such as "I have to go to work so I can eat," and then convert those ideas into outputs, our actions. And so information does flow through the human brain in this standard direction, known as bottom up, or feedforward.

But it also flows in reverse. When we look at an image, such as "My Wife and My Mother-in-Law," we don't simply convert the lines we see into faces. We also look for the lines in the faces stored in our memory. This is why you can see only one face at a time—either the wife or the mother-in-law.

Information in our brain doesn't just move from specific to abstract, as it does in calculators. It also moves backward, from abstract to specific. This reverse stream of information is called top down, or feedback.

View from the Top

What is the point of this top-down flow? It helps us navigate reality—more specifically, it helps us do it at the speed that we do.

We often forget how fast we are—we think of almost everything else, from plants to sea slugs, as slow and focus all our attention on an elite club of warm-blooded creatures—mammals and birds—which can produce distinct patterns of behavior on millisecond timescales: a virtuoso pianist playing an energetic piece; a falcon falling onto its prey from the sky like a dark lightning bolt.

From the standpoint of the rest of nature, we are terrifyingly fast. Our ability to steer our bodies through the chaos of experience at such speed is almost paranormal. It is simply impossible to gather complete information about the world at the rate at which we act.

So, instead of trying to perceive the world as completely and accurately as possible, we use what we already know about the situation to rebuild the missing parts, interpret the context, and cancel out alternatives and distractions. This is what the top-down flow of information does.

What we experience as sensations are not simply signals arriving into the brain from the sense organs. They are also, in equal part, interpretations of those signals imagined by the brain. Bottom-up and top-down flows work together to create a conscious percept. You can't separate one from the other. Perception is part imagination—that's what "My Wife and my Mother-in-Law" highlights, as do most visual illusions.

You can actually deduce the semi-illusory nature of perception without any clever drawings: simply consider the fact that we perceive depth. All our visual information comes from two flat retinas, neither of which has a way to distinguish a large object that is far away from a small object that is close—it's physically impossible. However, if you compare two flat images taken from different angles, you can make that distinction, because the position of the object nearby changes more than the position of the object far away. That's why we need two eyes looking at the same scene. The brain combines information coming from both eyes and produces a virtual three-dimensional percept that matches both two-dimensional inputs. In a 3D movie, this ability of our brain to synthesize two images is used to trick us into seeing depth where there is none. The movie is shot from two different cameras positioned at two different angles, and the two resulting videos are overlaid on top of one another and delivered into separate eyes using special 3D glasses, as if one eye was looking through the first camera and the other eye through the second.

You can use this conversion of two images into a single percept to do something startling called binocular rivalry. In 3D movies, the two images are only slightly different, and the brain can account for the difference by working out a single three-dimensional percept consistent with both of them. But in binocular rivalry, the two images are deliberately made incompatible: there's no percept that accounts for both. So the brain gets stuck in a loop: the moment it creates a percept, it starts conflicting with one of the eyes, and so the percept switches, but then it starts conflicting with the other eye.

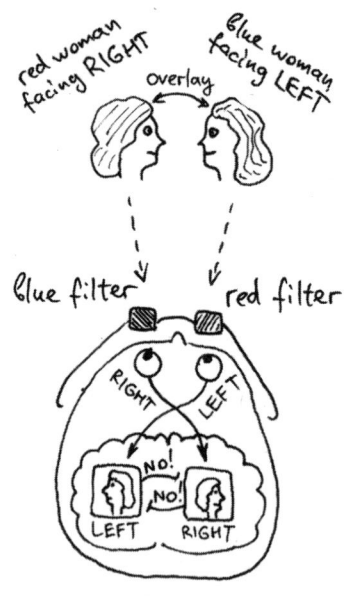

To do it in class, I bring a stack of cardboard 3D glasses with red and blue lenses and show students two overlaid images: a red picture of a woman facing left and a blue picture of a woman facing right. What most people initially see when they put on the glasses is only one of the images, maybe the one with the woman facing left. But if they keep staring at the screen, at some point the image gets blurry and switches to the other one, the woman facing right. A few seconds later it switches back to the original one, and so it goes indefinitely.

The reason it is so startling is because binocular rivalry feels like a glitch in your own brain's software—it exposes that our perception can be tampered with using some very simple tricks, like 3D glasses paired with incompatible images. It simultaneously exposes the physicality and the fragility of consciousness. It's like sticking a crowbar into a basic mental task that we rarely even notice.

We lean on a comfortable belief that our perception depends only on what happens outside of us—that we are somehow neutral observers of the world presented to us by our senses. But the existence of the top-down flow of information means that our minds are inherently subjective—that even as we experience something, we simultaneously make decisions about what it is that we are experiencing. It is a personal, subjective decision that depends on our memories, beliefs, and personalities, and not just on the outside world itself. If we can't make this decision—as in binocular rivalry—we get stuck in an endless loop.

It is no coincidence that with the advent of advanced digital computers, and especially the emergence of artificial intelligence, these human weaknesses—subjectivity, bias, confusion between reality and imagination—have also become the weaknesses of machines. It is precisely the combination of bottom-up and top-down information flows that powers artificial neural networks—the software that underlies today's chatbots and image generators. Its very similarity to the human brain is what gives AI its power and its weaknesses, making it so humanlike in every respect.

So human perception is distinct from any neural activity that goes on in a brain of any animal. It is not simply abstraction of outside information that terminates in some specific result. It is a constant dialogue between the world and the brain, a negotiation between what is really happening and what we believe should be happening.

What follows from this is equally profound and unsettling. Our brains, it turns out, are not even trying to determine what the world really *is*. Instead, what they are trying to do, with their two opposing flows of information, is *align reality with expectation*.

Perception as a Controlled Hallucination

It certainly *feels* like we see the world for what it really is.

That's what makes "The Dress," a photo that went viral in 2015, so bewildering. For some reason that is not entirely clear, the famous image borders two possible interpretations: a white-and-gold color scheme and a blue-and-black combo. At least part of it seems to do with how one interprets ambient light in the photo: those who imagine the picture taken in daylight gravitate toward perceiving a white-and-gold colored dress, whereas those who picture it as artificially lit tend toward blue and black.* But what's most striking about this image is the very fact that different people in the same room can look at the same image on the same screen and see it so differently.

It would not be surprising if different people had different beliefs—for example, if two people read the same piece of news but one got upset and another happy. But we are not used to disagreeing about basic properties of objects, like color or shape. Generally speaking, we all arrive at consistent interpretations of those properties—we don't argue, for example, whether a soccer ball is a sphere or a cube. So we assume that these properties exist outside of us, independently of our knowledge and memory, and that when we see them, we perceive the world for what it really is. And yet "The Dress" shows that it's not the case—the same photo on the same screens can lead to two different "what it really is" interpretations. Which makes you realize that the sensation of solid, tangible reality is an illusion—as malleable and individualized as a political opinion. We just happen to have consistent opinions about colors and shapes in most cases. If you fully process the implications of this, it can feel like the ground is shifting beneath your feet.

We are, in fact, so convinced that we experience the world fully, objectively, and unambiguously that it is hard to believe how messy and incomplete our perception actually is when we test it.

What would you say is the angle of vision within which objects appear clear and sharp? In my experience, most people tend to place it somewhere around 45 degrees, if not 90 degrees. In fact, the maximum-resolution area of our retina corresponds to a visual field of only 2 to 3 degrees—about the width of a thumbnail of an outstretched hand. Compared to this small area called the

---

*This, in turn, correlates with being a morning lark or a night owl.

fovea, the rest of the retina is very bad at discerning anything. At the edge of an actually quite narrow visual field of 20 degrees, visual acuity drops tenfold. This is enough to notice when something happens in your peripheral vision, but generally not enough to figure out what it is. Just look at the center of a page and try reading it from the top without moving your eyes.[1]

Despite this glaring deficiency, people are confident about what they see in their peripheral vision. If you repeatedly flash one image in the corner of a person's eye but then quickly change it to a different one each time the person moves their eyes to look at the image more closely, eventually the person will believe that they saw the second image even when they were shown only the first.[2]

How, then, do we manage to see anything meaningful at all? First, we constantly move our eyes—on average, three to four times per second. But more importantly, we don't experience a live video stream from our eyes—such a video would look nauseatingly jerky, blurry, and disconnected whenever our eyes moved. Instead, we inhabit a hallucination, a smoothened virtual reality constructed from our memory, which gets rapidly updated by our eyes' rapid snapshots of the environment.[3]

It just doesn't *feel* this way. It always feels as if we are perceiving everything there is to perceive, and that this perception simply descends on us without us having anything to do with it. It's easy to challenge this illusion. For an assignment, one of my students did this: at random times during the day, she would freeze and try to write down every object she remembered on her right side. She found that her recollection was astoundingly incomplete: she could only recall things she had previous interactions with—she would remember, for example, a trash can she had once used but missed the one she hadn't previously engaged with, even though it was right there by her side.

So our brains are really doing a number on us: they actively sculpt information coming from our sense organs into a vivid hallucination, and then not only convince us that this hallucination is what the world really is, but that it's *all* that it really is.

Of course, this is exactly what nature intended. We don't need to be aware of everything. We only need to be aware of things that we care about—that perturb our lives in some way. The rest is distraction that should be canceled out. We inhabit a predictable, cohesive virtual world constructed around those things we care about—a model of the world as it is known from our point of view. New information arrives into this virtual world from the bottom up but is selected, filtered, interpreted, and parsed into concepts and objects from the top down. Think of a cocktail party: there's a wall of sound, but suddenly you hear your name coming from the far corner of the room. In the same way, we constantly pick out relevant things from the environment and ignore others that we deem irrelevant—except that we are mostly unaware that this is happening and genuinely believe that our cherry-picked elements of experience represent the world as it truly exists. This is why psychedelic drugs, which, simply put, ramp up the bottom-up stream of information,[4] feel so *revelatory*—they show you that there's infinitely more to perceive than you realize most of the time.

In the last chapter, we talked about abstraction and memory, the core operations of any brain, from sea slugs to humans, and then saw how they were embodied in the brain's most basic building blocks—neurons and synapses, elements that we share with even our most primitive animal ancestors. We are now discussing something that's built upon those elements but is vastly more complex—human perception, which combines two opposing streams of information.

This dual essence of human perception, as I just described it, is also embodied in one specific unit of our brain—but this one is not a simple neuron or synapse, but rather our brain's most advanced element, its crown jewel, and arguably the most wonderful thing nature has ever created: the cerebral cortex.

The Thinking Surface

Why exactly does the human brain look like a walnut? That doesn't help to increase the number of neurons—the volume of the skull stays the same regardless of all the sulci and gyri. But you *can* increase the surface area: the more folds, the more surface per unit of space inside the head. It is this surface—a few millimeters of cerebral cortex, or gray matter, coating our brain on all sides, that is so precious. If you look at the cross-section of a human brain, you can almost see the hand of evolution jamming those folds into our skull.*

Look up a video called "A causal test of face recognition in the FFA." In the video, you can see a patient sitting in a hospital bed, his head covered in bandages that hide electrodes implanted into his cortex. The reason for this procedure is to treat epilepsy, which sometimes requires implantation of electrodes in order to locate the damaged region of the brain. But the procedure also presents a unique window into human consciousness. Patients must remain awake to report the sensations during stimulation, and so they can also tell you what exactly happens in their minds when specific locations on the surface of their brains are triggered.

The patient in the video has electrodes positioned to stimulate the fusiform face area, a location in the cerebral cortex known to be responsible for the perception of faces. The doctor can be heard speaking in the background. "Look at my face," he says, as he counts down to a sham stimulus to avoid any

---

*A much more puzzling question, incidentally, is the reverse: why do walnuts look like brains? It might be a complete accident. But it's possible that there is some functional reason for the shape. If so, it must have something to do with the increased surface-to-volume ratio.

possibility of a placebo effect: a button is pressed, the stimulator clicks, but no stimulus is delivered. The patient remains indifferent. Then, another countdown, another click, and this time a real stimulus. Immediately, without hesitation, the patient says, "You just turned into somebody else. Your face metamorphosed." The video progresses through a series of sham and real stimulations, and each time a pulse is delivered to his brain, the patient very calmly and clearly describes how the doctor's face simply melts, metamorphoses, shifts, and stretches, somehow without becoming larger or smaller—"it was more of a perception, how I *perceive* your face," says the patient, who is lost for words when prompted to explain further.

The video is startling for the same reason binocular rivalry is startling: it simultaneously exposes the physicality of the mind and its fragility. A single electrical impulse delivered to a specific location in the brain is sufficient to induce a predictable hallucination that, in the ancient times, could have had only mystical, paranormal explanations. But what makes it even more eerie is the fact that the patient describing this hallucination does it so clearly, without a shadow of hesitation or confusion. Had he been incoherent, the video would not have such an effect. The fact that you can be perfectly alert, logical, and unintoxicated and yet experience this complete breakdown of a very specific mental capacity shows that our minds are not monoliths—they consist of many small pieces that are tightly woven together yet can be teased apart.

This technique of stimulating different areas of the cortex in order to find the problematic location that causes epileptic seizures—somewhat reminiscent of the Windows game *Minesweeper*—was pioneered in the 1930s by the American neurosurgeon Wilder Penfield. Eventually he and others who used his method produced an entire map of the human cortex, as if it were some extremely complicated video game in which each button interaction corresponds to a sensation or motion: *this location causes a twitch in the left middle finger, this one triggers a smell of roses, and that one induces the feeling of déjà vu.*[5] It turned out that not only is our mind subdivided into all these small pieces that can be separated and triggered independently, but all these pieces also have a specific location, a dedicated address somewhere on the surface of the cortex that is consistent among different people—such as the fusiform face area, the "face" region that was stimulated in the patient on the video.

Wilder Penfield electrically stimulates various areas of the cerebral cortex

Patient reports sensations

A map of correlations is established

In the absence of electrodes implanted into the brain, what triggers all these different parts of the cortex in real life? Experience does: all the sensory signals brought in by eyes, ears, and other sense organs.

But we already established that this ongoing experience is too fast to be meaningfully analyzed in the moment: there's not enough information to press all the correct "buttons" every time.

What happens instead is the "buttons" themselves *actively compete with one another* to be "pressed" by the meager, incomplete shreds of information. It is exactly the kind of competition that binocular rivalry exposes by trapping the brain in an endless loop. This competition between the "buttons" of our cortex is the core process that underlies our conscious experience, from seeing depth to believing in God.

An Array of Columns

What I figuratively referred to as "buttons" in the cortex corresponds not just to individual neurons, but more precisely to cortical columns. A cortical column is a standardized module of about a hundred interconnected neurons that traverses the cortical surface. The flat sheet of the cortex is formed by millions of these cortical columns stacked next to each other. A cortical column acts as a unit—when it's activated, all of its neurons are activated together.* So you can think of your consciousness as a musical piece played on this keyboard of cortical columns, and the *Minesweeper*-style experiments on epileptic patients as pressing single keys on the keyboard using implanted electrodes.

Each column contains different types of neurons. So far we have discussed only the more typical, excitatory kind: when such a neuron fires, that next neuron down the line receives a small injection of positive current, which brings it closer to firing (excitation) itself. But there are also inhibitory neurons, which do the opposite: when they fire, they inject into their targets not positive, but negative charges, taking them further away from excitation. Each cortical column has some excitatory and some inhibitory neurons that are plugged into other columns. So when a given cortical column becomes excited, it delivers an excitatory boost to its "friends" and an inhibitory blow to its "enemies." It's as if each key on the musical instrument not only triggered its own note, but also simultaneously triggered other keys and their notes and blocked yet others. This is what happens when new information arrives at the cortex: cortical columns start frantically activating and inhibiting each other, and eventually a stable arrangement emerges: a small, specific minority of columns that activate each other and inhibit everyone else, producing a continuous, specific "chord." That's when the brain settles into a percept.

For example, when brushing my teeth recently, I noticed a funny visual effect: I looked at a tube of moisturizer from far away and thought I saw the word "RICE," but when I moved closer, it switched into "ALOE." The skinny graphic font made *A* and *R*, *I* and *L*, and *C* and *O* look similar, and whenever the word was far away and blurry, it stubbornly clicked into "RICE" even after I realized the mistake. I could pinpoint the exact location where the words switched. So it would appear that somewhere in my brain, there are cortical columns that represent the word *aloe*, some that represent the word *rice*, as well as columns that represent the individual letters. The letters and the

---

*Actually, the reality is messier, because sometimes only part of a cortical column can be activated, which creates a further layer of complexity. But we will ignore this nuance for the sake of the general picture.

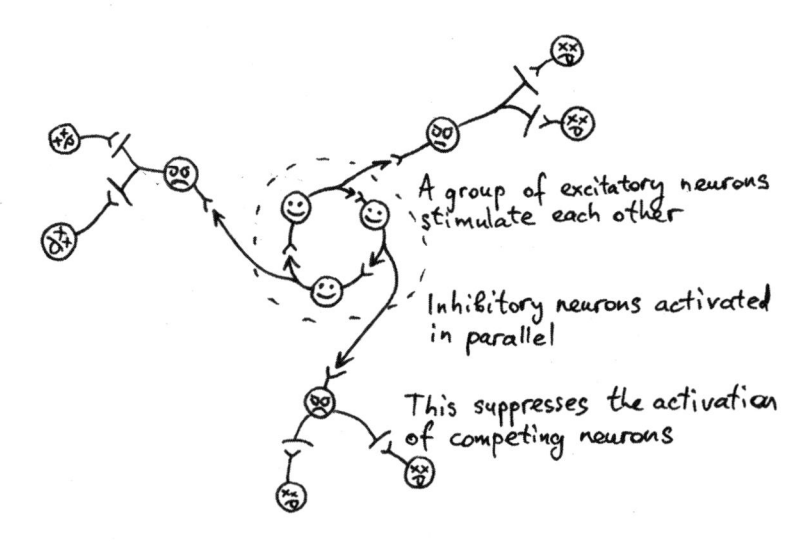

A group of excitatory neurons stimulate each other

Inhibitory neurons activated in parallel

This suppresses the activation of competing neurons

words coactivate each other, forming a cohesive ensemble, bouncing electrical excitation among themselves in a loop. When I look at the text closely, sensory information flows into *A, L, O,* and *aloe,* and so those columns power each other up, while simultaneously inhibiting *R, I, C,* and *rice.* When I step far away—the power of *A, L, O,* and *aloe* diminishes, and *R, I, C,* and *rice* emerge instead, maybe because I like rice so much and so these columns hold more sway in the cortical sheet when clear sensory information is lacking. It's a winner-take-all competition between different ensembles of cortical columns. Whoever wins gets to remain "on" and keep circulating excitation.

This circulating excitation among select columns, by the way, is how we hold objects in our mind. If it continues circulating after the sensory information is removed, we call it working memory.[6] For as long as a particular ensemble can hold their specific configuration, you remember what this configuration represents, and if the ensemble dissolves or gets outcompeted by other columns, you forget.

How the Model Forms

How do these columns come to represent different things in the first place? How is *aloe* assigned to some columns, and *rice* to others, and how do these columns connect to the letters of the English alphabet?

The answer is synaptic plasticity, the essential ability of brain cells to adjust their wiring in response to the signals that pass through them.

Genes set up the overall structure of our brain but leave it to real-life experience to figure out all the fine-tuning. The process begins before we are even born: astoundingly, a developing eye sends waves of signals to the developing cortex, almost like a test video, and the cortex uses this mock visual signal to set up its visual areas.[7] If the signal is absent, the cortex doesn't even know that eyes exist: it needs live experience to train itself to interpret it. This also means that in theory, you could implant a totally different sense organ into the cortex, which will set it up just like a real one—this has actually been done in rats, which had infrared cameras plugged into the cortex through an array of electrodes, and the rats eventually learned to use them together with real eyes.[8]

In early childhood, neurons in the cortex grow a massive number of connections—way more than needed. Then, for an *extremely long period of time*—until the late twenties for most people—the connections are pruned, leaving behind only the ones that best predict and explain ongoing experience. We learn which objects are associated with which properties, which words, which letters. Useful connections remain; others disappear. The configuration that remains *is* the meaning of *aloe* and *rice*. There is no further "aloeness" or "riceness" to columns that represent these concepts. The very concept of *aloe* is contained in the fact that it is linked to letters *A*, *L*, *O*, and *E*, as well as to the image of an aloe, to the texture of the plant's flesh, to the abstract idea of a hand moisturizer, and to every other experience we have had with aloe. Each of those elements—such as the letter *A*—is, in turn, defined by its relation to yet other elements: the image of two slanted lines with a horizontal line in the middle, the sound *a*, the words containing the letter. A particular set of cortical columns comes to represent *aloe* simply by standing at the crossroads of all these strands of experience related to aloe.

In other words, the cortex doesn't simply assign different parts of the external world to different neurons or cortical columns. Rather, it uses its own internal architecture to re-create relationships between elements of the external world.

It's like the *Oxford English Dictionary*. It doesn't show you a photo of a bird for the word "bird" or play a sound for the word "sound." It is a book written in English words that defines every English word through other English words. It represents the English language as a self-contained set of relationships among elements. And in the same way our cortex represents the external world and all of its objects and processes as a self-contained set of relationships among cortical columns.

This relational map of the world, its basic breakdown into mental objects and categories, gradually solidifies, becoming more restricted and less ambiguous as we grow older. This is why we refer to our youth as formative years—it is literally the time when our pathways of thinking are formed.

But the cortex never becomes completely rigid. It continues to change throughout our life, both in the short term and in the long term. Typing on a smartphone, for example, gives more power to the areas representing thumbs—the signals they send to other parts of the cortex become stronger.[9] The extra oomph gradually decreases over several hours. In skilled musicians, however, the areas representing fingertips on the playing hand retain stable power: they are physically larger and more "influential."[10]

Such long-term changes in the cortex are much harder to achieve in adults than in children. There is, however, a shortcut: an appendage of the cortex called the hippocampus. It is the structure that enables us to memorize things not by rote repetition, but instantly. The hippocampus is highly plastic. It essentially takes a snapshot of a state of the cortex at a given moment and stores it in its own configuration, like a hyperlink to an experience. Later on, the hippocampus can reactivate that same state in the cortex, taking us back to an episode from the past. This is what we mostly imagine when we think of a memory in general. Actually, it's a very specific type of memory, called an episodic memory—a state of mind canned by the hippocampus and reheated by the cortex. If a memory is replayed like this over and over, it eventually changes the cortex itself—that's how a single memorable episode, such as a chance encounter or a traumatic event, can gradually work its way into our very patterns of thinking.

Planting the Tree

To summarize: in the cortex is a field in which cortical columns representing different aspects of the world actively compete with each other to explain what is going on. How do the top-down and bottom-up streams of our perception fit into this competition?

When Wilder Penfield started his "*Minesweeper*" experiments in the 1930s, he quickly realized that the cortical "buttons" inducing different sensations were not randomly scattered across the cortical surface. They are organized—in a strange, counterintuitive way—on a map of the cortex that can be said to be the map of our mind.

The top-down and bottom-up streams travel in opposite directions across this map.

Here's how the map is organized. First, there are large areas corresponding to different senses. Vision is represented in the back part of the cortex; hearing in the sides; touch (somatosensory cortex) runs across the top of the brain like a thin band. There are also vast areas of cortex unassigned to specific senses, sometimes called associative areas—more on them later. Motion has its own region, the motor cortex, which looks very similar to that somatosensory cortex band, and runs as another band right next to it.

Then, within each area, there are subareas corresponding to different levels of abstraction.

This needs some unpacking. Let's say you are looking at an image of a house, a simple triangle over a square. Information from your eyes is relayed through several brain parts and then enters your cortex in the subarea called V1, or primary visual cortex, where it activates an assortment of columns. There are not only columns for every part of the visual field, but many columns per location for every possible thing the eye could be seeing there—that's a lot of columns, and this subarea is very large. Some columns represent a black dot slightly left of center, and some represent a green dot in the top right corner. Each time your eyes look at something different, a different set of columns activates. So the image in the eye gets transformed into a pattern of activated V1 columns.

Next, each of these V1 columns sends a signal to the next subarea, V2. It's a much smaller area—many V1 columns converge onto individual V2 columns. Here, this input from V1 activates a new assortment of columns. But these columns in V2 now correspond not to different kinds of dots, as in V1, but to different kinds of *lines*, each kind of line being a particular pattern of dots. There could be, for example, a column in V2 representing *left-tilted line in the top left corner*, and another representing *horizontal line in the middle*, each of them activated by a particular combination of "dot" columns in V1. So now the image is represented through more abstract entities—it's similar to the difference between a bitmap and a vector image.

The process is repeated over and over, from V2 to V3 to V4 and so forth (in reality, there are multiple parallel pathways, and the neat numbering soon

[Figure: Hand-drawn diagram showing a brain-like shape with labels V1, V2, V3... and progression from dots/lines to shapes (triangle, square, circle) to houses and a person with shovel. Labels: "more granular representations" on left, "more abstract representations" on right. Arrows showing "top-down" (leftward dashed) and "bottom-up" (rightward solid). Caption: "Signal flow through the cerebral cortex"]

breaks down). In each successive subarea, representations become more and more abstract. Various combinations of "line columns" activate columns representing shapes—*triangle* or *square*. Those activate further columns representing objects—*house* or *shovel*. Those activate even further columns corresponding to types of objects—*buildings* or *tools*, for example. Information progresses through the cortex from level to level, from more specific to more general, from more granular to more abstract.

There's some room for confusion in terms here. In everyday language, the word "abstract" most often comes up in reference to abstract art, and that art tends to look more like lines and shapes than houses and shovels—think of Kazimir Malevich's *Black Square*. So why am I saying that a column that represents *house* means something more abstract than a column that represents *square* or *triangle*?

By definition, the word "abstract" means "dissociated from specifics." When we are talking about the brain, the specifics are the granular elements of experience that the sense organs detect at any given point—for example, the color of the line that forms the base of the drawing of the house or the angle at which the lines forming the roof intersect. The abstraction, in this case, is the *idea* of a house—this idea doesn't change whether you are looking at the top of the image, at the bottom of an image, at a drawing, or at a real house. So a shape is lower in its level of abstraction than the identity of the object is.

With abstract art, the word "abstract" refers not to the actual black square on the canvas, but to a complex artistic idea that it represents—I think a lot of people miss that nuance. Malevich's *Black Square* is a companion piece to "Manifesto of Suprematism," a philosophical treatise that describes a full theory of liberation from state, religion, and objective representation of nature in favor of "supremacy of pure feeling." So, in art, the specifics are the traditional depictions of people, objects, and landscapes, and the abstractions are the bigger ideas that abstract artists set out to represent. It just so happens that they often represent them using simple lines and shapes, which, for the brain, are lower-order visual abstractions, from which higher-order abstractions—*house* and *shovel*—are built.

Senses other than vision work similarly to this assembly of *house*, the abstract idea, from granular dots and lines. When we hear a sound, it first activates combinations of cortical columns corresponding to various frequencies (granular); those then activate columns corresponding to patterns of frequencies such as timbres and notes (more abstract), those activate columns for patterns of patterns such as songs and artists (even more abstract), and so on. Touch, taste, smell—they all begin with granular representations, which then progress through levels of abstraction.*

At some point—at a certain level of abstraction and in a certain location on the cortical map—different senses converge. An abstract representation of a guitar, for example, might get activated any time you see, hear, or even touch a guitar, induced by signals emanating from any of those senses. Those unassigned regions of the cortex, which represent objects, abstract concepts, ideas, and beliefs, are collectively called "associative cortex"—as in, those are the parts of the cortex that associate different senses with each other. A better term for them, in my opinion, would be "abstract cortex": they are the parts where information reaches such a level of abstraction that it no longer belongs to just the eyes or the ears alone.

---

*However, smell—olfaction—is somewhat unique and arrives into the cortex through its ancient appendages called olfactory bulbs.

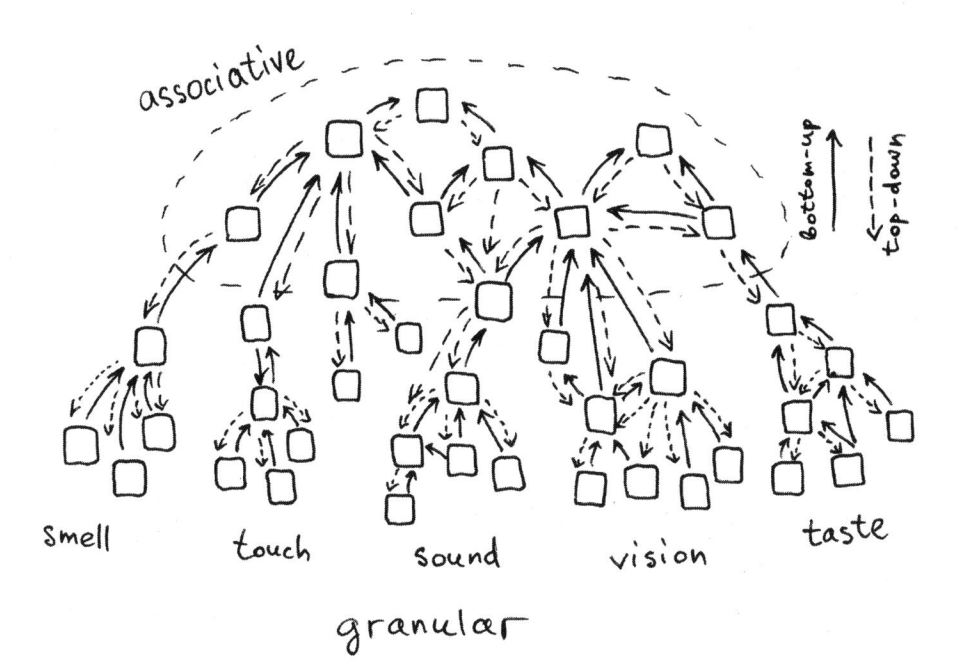

Map of the cerebral cortex (labels: abstract, associative, bottom-up, top-down, smell, touch, sound, vision, taste, granular)

In the previous chapter, we discussed an "inverted tree" of abstractions, a space that our mind inhabits. At that point we talked about abstraction in general and thought about the tree in purely introspective terms. Now we have a place to put it, to embody it in the physical reality of the brain. The tree is the network of hierarchical associations between cortical columns, progressing from specifics to abstractions, from the branches of the tree to its trunk, from sensory areas to associative areas of the cortex. You could also call this network a fractal, a geometric figure consisting of many repeated levels.*

So far I've described the bottom-up flow of information, the progression of signals from the branches of the inverted tree to its trunk: from eyes, to dots, to lines, to *house*, to *building*. But there's also the top-down flow, from the trunk

---

*I cannot resist sharing one of my favorite jokes about the famous theorist of fractals, Benoit B. Mandelbrot: what does the B. in "Benoit B. Mandelbrot" stand for? "Benoit B. Mandelbrot."

Fire from Within **199**

to the branches: from *building*, to *house*, to lines, to dots. It operates at the same time as the bottom-up flow.

Any time a cortical column becomes activated, it passes information to the level above: for example, *square* and *triangle* converge on the higher-order column *house*. At the same time, they each pass information to the level below: both *square* and *triangle* project their influence onto the columns representing individual lines, helping them outcompete other, irrelevant lines, like creases on the paper.

It is as if the bottom-up flow is saying: here are these *dots* that we are looking at, and since they are arranged the way they are, we must be looking at some vertical, horizontal, and tilted *lines*. And since there are such *lines*, there must be a *triangle* and a *square*. And since there is a *triangle* and a *square*, this must be a *house*.

At the same time, the top-down flow is saying: since there is a *house*, there must be a *triangle* and a *square*. Since there is a *triangle* and a *square*, there must be vertical and horizontal and tilted *lines*. And since there are such *lines*, there must be such and such *dots*.

These two conversations are happening in parallel between the same cortical columns—*house*, *triangle*, *square*, a specific set of lines, and a specific set of dots. Each of them sends a signal to the level above and a signal to the level below. This two-way communication simultaneously helps us interpret sensory information and actively sculpt what we are seeing.

Here's an example of how this two-way motion of signals up and down the tree of abstraction manifests in everyday life. The standard phrase for "queuing" in American English is "to get in line," but in New York it is "to get on line." For most Americans, the "line" refers to the people standing one behind another. For New Yorkers, it instead refers to an imaginary line on the floor. I had been living in New York for four years before someone told me about this, and all that time I had heard only "get *in* line." It simply did not occur to me that there could be any other phrasing—I had no other top-down pathway to interpret the sounds I was hearing, and so I always heard them the way my top-down pathway predicted: "*in* line." But then, when I learned about it, I formed that second pathway, and now I can use both—"in line" and "on line"—to hear two different things. I delight in telling new New Yorkers about this, because it's such a consistent and vivid example of how expectation shapes perception, and how new information can make you see (and hear) the world in a new way.

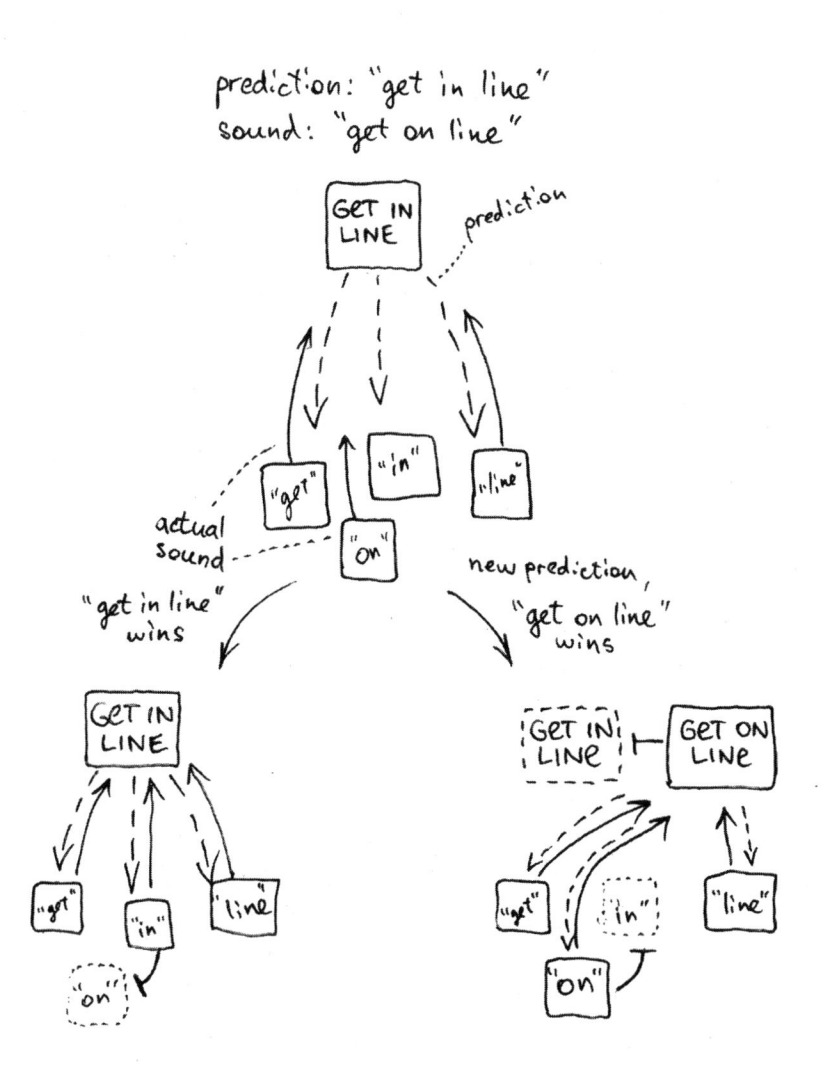

Any time we perceive something, information spills over the cortex, runs up the inverted tree of abstraction, and starts circulating through it, moving up and down in waves across huge numbers of cortical columns across many levels of meaning. The specifics of every new experience are filtered through already existing abstractions to initiate new abstractions. Those go on to initiate further abstractions, while at the same time changing the specifics. And so signals continue to bounce around the cortex, fueled by new infusions of experience, at every waking moment of our life. It never really stops. All levels are jostling to come to a consensus—a stably circulating, minimum-effort wave that binds them together. But they never truly do—and the result of this Sisyphean quest for peace is our conscious experience.

Consciousness is not any particular result of information processing in the brain—but rather, consciousness is its very flow, its rolling motion from the past to the future up and down the inverted tree of abstraction, the unstoppable wave of information bound by cause and effect across levels of meaning.

Theories of Consciousness

The functioning of the cerebral cortex I described here is consistent with two of today's most prominent theories of consciousness: integrated information theory (IIT) and hierarchical predictive coding theory.

The two theories approach consciousness from two different angles. IIT looks at it as a property of software: it says that information becomes conscious in a specific configuration in which highly integrated (very abstract information) has a causal effect on whatever information it is integrating (less abstract information) and vice versa in a circulating loop of integration and causality.[11] This accurately describes how information moves across the cortical hierarchy up and down the tree of abstraction. There are columns somewhere in the cortex that represent extremely integrated, abstract pieces of information, such as *myself*. These peaks of the hierarchy project top-down signals onto the rest of the information arriving in the cortex and vice versa; all those pieces of information have a bottom-up effect on how this *myself* operates.

Hierarchical predictive coding theory looks at consciousness more like a property of hardware, a feature of our brains. It says that the cortex stores a model of the world in the arrangement of its cortical columns: it *predicts* what sensory inputs should arrive in what situation. What we consciously perceive is not the model itself, but rather deviations from the model—the parts that weren't predicted, interpreted, and explained. Top-down signals—the ones sourced from within the cortex—represent predictions. Bottom-up signals—incoming data from the sense organs—represent data to be interpreted. If all data fit the predictions, there's no competition between columns, and the cortex remains at peace. Only if something is left unexplained does competition ensue, leading to conscious awareness.[12]

This is why we don't notice things that don't actively challenge our awareness. There's a remarkable video I show in class (watch it now: look up "Transport for London awareness test"; note: spoilers follow) in which viewers are asked to count the number of passes a team of basketball players makes to each other, and as the viewers count, a person in a bear costume walks backward across the screen among the players. I have shown this video to hundreds of students, and there've ever been only one or two laughs when the bear first comes out. "It's easy to miss something you're not looking for," explains an onscreen caption. "Watch out for cyclists."

According to the hierarchical predictive coding theory, what our cortex wants from reality is a perfect fit: all senses perfectly predicted. So it is totally unfazed by sensory inputs that don't challenge the fit, such as the ones we are not paying attention to. In fact, you could say that attention *is* the vigilance with which we detect mismatch between a particular aspect of experience and our prediction of it. Not paying attention means not actively detecting errors.

When there's a mismatch between senses and expectations, two things could happen.

The first outcome: the bottom-up stream wins, and the top-down stream changes its course—we change our mind, adjust our expectations, notice something we haven't noticed before. That's what happened when I learned about "getting on line" and started hearing it.

Second outcome: the top-down stream wins. You stubbornly perceive what you expect to perceive, regardless of evidence (as when I kept hearing "get in line," when New Yorkers say "get on line," or when the word *ALOE* looked like *RICE*)—which is to say, you hallucinate. Or, even more stubbornly, you get up and change the world, forcing internal beliefs onto the outside world through deliberate action.

In this context—hallucination versus physical action as two forms of the top-down stream "winning"—it is especially interesting to have a look at the motor cortex, the part that sends outputs to the muscles.

In general, bottom-up and top-down signals travel across the cortical map through distinct layers, so they don't mix together. There are six layers in most parts of the human cortex. Very roughly, you could say that bottom-up signals enter a given cortical column mostly through the central layer 4, and top-down signals through superficial layer 1.[13] And here's the kicker: *the motor cortex does not have a layer 4.*[14] It has other layers, but not the one into which you plug in the bottom-up signals. This is the main difference between a motor cortex and a sensory cortex: columns in the visual cortex, for example, receive both top-down and bottom-up signals, whereas columns in the motor cortex are only controlled from the top down.*[15]

This is as weird an idea to wrap your mind around as it gets: to the cortex, *motion is like another sense, but one that is engaged only from the top down and not from the bottom up.* Vision, hearing, touch—they all come from the outside *and* from the inside. Motion only comes from the inside, but otherwise it works the same: just as we use our memory and imagination to shape an image into one pattern or another—either "wife" or "mother-in-law"—we use our memory and imagination to shape our clothes into a familiar pattern before going outside and to shape our apartment into a pattern we deem tidy enough. To the cortex, it doesn't matter whether the familiar patterns occur inside of it or outside of it: it enforces them through top-down signals regardless of whether that's done through perception or through action. Action, in other words, is *embodied hallucination.*

The two theories of consciousness—IIT and hierarchical predictive coding—are really talking about the same thing. IIT is mostly interested in the question of why reality *feels* a certain way, and its answer is: because levels of highly packaged information constantly update each other, and that's what feeling something is. The predictive coding theory is more interested in why we *act* a certain way, and the answer is: because we strive to match reality with expectation, and both our perceptions and actions are, in fact, the very process of matching.

Basically, IIT describes the software that runs on hierarchical predictive hardware.

What lies at the heart of both these theories is the idea that consciousness is not something that is calculated, produced, and delivered somewhere where it achieves its final, conscious state. Rather, it is the rolling motion of information circulating through the brain, perpetually selecting some cortical columns over others, and each time prompting a reevaluation of the entire inverted tree of connections between them. The rolling motion is achieved by a combination of two signal flows: from specific to abstract and from abstract to specific.

---

*There are some conflicting reports on this account: some claim that the primate motor cortex does in fact have a layer 4, but if so, it is still much less pronounced than in sensory cortices.

Two Fires

It is as hard for us moderns to make sense of the way ancient philosophers thought about the human mind as it would no doubt be for them to make sense of ours. Consider a quote from *Timaeus*, one of Plato's dialogues, in which he describes how the outside world connects to the human inner world—in this case, through the eyes.

> And of the organs they [the gods] first contrived the eyes to give light, and the principle according to which they were inserted was as follows: So much of fire as would not burn, but gave a gentle light, they formed into a substance akin to the light of every-day life; and the pure fire which is within us and related thereto they made to flow through the eyes in a stream smooth and dense, compressing the whole eye, and especially the center part, so that it kept out everything of a coarser nature, and allowed to pass only this pure element.
>
> When the light of day surrounds the stream of vision, then like falls upon like, and they coalesce, and one body is formed by natural affinity in the line of vision, wherever the light that falls from within meets with an external object. And the whole stream of vision, being similarly affected in virtue of similarity, diffuses the motions of what it touches or what touches it over the whole body, until they reach the soul, causing that perception which we call sight.[16]

So to review: the body, according to Plato, contains some inner light, "pure fire which is within us," and this fire flows from our eyes to the outside "in a stream smooth and dense." When this fire from within meets the light of day, the two coalesce into a single body, which then transfers its motions to the soul. Vision is the collision of outward fire with inward fire.

Today this basic theory of vision is referred to as extramission theory, meaning that something is emitted from our eyes, and that is what makes us see. This theory held sway in the ancient world for centuries. Galen described it in similar terms almost five hundred years after Plato.[17]

To a modern science historian, it is little but an ancient misconception. Credit for overturning it is generally given to Ibn al-Haytham, a tenth-century Arab scientist also known by the latinized name Alhazen. In his seven-volume *Book of Optics*, he systematically annihilated the extramission theory with scrupulous precision worthy of today's scientific journals. Vision, he asserted, works in the opposite way: rather than radiating from the eyes, light enters the eyes from the outside. Ibn al-Haytham's key argument was that even if the eyes did "emit" some sensing medium, whatever it was, it had to eventually reenter the eye in some way. Therefore, the "emission" did not add anything to the explanation, and since it was not based on anything but conjecture, it could be disregarded. Several centuries later, this logical maneuver would become known as Occam's razor, in reference to an English Franciscan friar who would use it to similarly "shave off" anything superfluous to his reasoning.

The optics of vision were not definitively worked out until the sixteenth century, when Johannes Kepler showed that the eye's lens projects an image onto its inner back surface, the retina. (The ancients believed that the eye sees using its entire volume.) So unless you are a superhero, there isn't anything that your eyes radiate into the outside world.

Today, the extramission theory might seem wild—what were all those ancient philosophers smoking? For example, how do we see stars, if fire from our eyes has to reach them first?

Plato

Ibn al-Haytham disproves extramission

In Plato's time, though, the theory was not at all unreasonable. Without special equipment and knowledge accumulated over millennia, it is not obvious that stars are very far away or that light has to travel anywhere.

At the same time, extramission appears to be a surprisingly intuitive theory, despite the fact that it's not how vision really works.[18] Children, unadulterated by scientific knowledge, seem to generally agree that eyes project something into the outside world. This intuition is even reflected in language: we *look at* or even *lay our eyes on* objects, rather than the objects looking at us, as the wind blows at us, or water falls on us.

Maybe the extramission theory is not as ridiculous as it first appears. It just needs to be reframed in modern terms. Today, vision is the subject of neuroscience, the science of the brain. In the ancient world, vision was the subject of optics—the science of light and its interaction with the eye.[19] In the premodern world, sensations were generally agreed to be produced by sense organs such as eyes and ears, which communicated them directly to the soul—there was simply no need for the brain to be involved in this. Consequently, the ancients were almost farcically uninterested in it. Some scholars considered it a cooling device; others a sperm-producing gland. There was a theory in ancient Egypt that the brain's function was to deliver mucus to the nose.

The brain is what Plato missed in his theory of vision: he concentrated on the eyes, but eyes are in fact merely the brain's peripheral, and largely passive, devices. It is the brain that makes vision an active process.

And if we do let go of the eyes and instead use Plato's metaphor to understand how the brain generates perception, the ancient Greek account of "two fires" colliding from two directions in a unified, undulating body falls stunningly in line with modern science. According to Plato, vision is the fusion of two types of fire: one radiated by objects and another, a "fire from within," emanating from the eyes. According to modern neuroscience, vision is the fusion of two streams of signals: one, bottom up, originating in the outside world, and the other, top down, flowing from within the brain.

The only difference is whether the two fires fuse before the eyes or behind them.

Chapter 11.
The Dark Room

Is living really worth the effort? And yet you keep on living—out of curiosity: waiting for something new.
MIKHAIL LERMONTOV

A brain is an amazingly useful thing.

It helped our jellyfish ancestors swim using muscle contractions. It helps sea slugs respond to abstract ideas, such as danger. It helps us zoom through the chaotic world of our senses without even noticing something's amiss.

But it seems that something's broken in our relationship with our brain.

As humans, we often feel that we are at war with ourselves. We want what we cannot have and need what we don't want. We get addicted to bad things and lose interest in good things. We ruminate, we obsess, we snap, we regret. It is as if we are always trying to get to some fuller, better, more complete, more natural version of our lives and never quite get there.

Why are we so misaligned with our own brains?

The temptation is to assume that our lives as modern humans are unnatural, and so it prevents us from realizing some primeval happiness that our ancestors presumably all shared. The cavemen had no French fries, so they didn't have to worry about obesity or force themselves to go to the gym. They spent their days blissfully walking in the woods gathering nuts and berries with plenty of fiber. They had no money or jobs or marriage or religion or drugs, so there was no inequality or violence or jealousy or hierarchy or addiction. It is only when we abandoned this hunter-gatherer paradise for the temptations of agriculture and civilization that our lives became so discordant with our biological needs.

Of course, this vision of a carefree past is not actually true. We don't know much about the psychology of our hunter-gatherer forebears, but there's one thing we can be sure about: they were just as grumpy and restless as we are.

Our frustration with life is nothing new. It is there by design—a design that runs much deeper than civilization, deeper even than the human species. It is this design that keeps us perpetually aggravated, teasing us, prodding us, like a voice from an ancient, animalistic past that whispers into our ear: *there's more to life than what you have.*

We are not meant to feel satisfied by what we have. We are meant to look for more.

Out of the Dark Room

We already talked about how the brain decides what to do. In the previous chapter, we discussed the cerebral cortex, our brain's universal machine of understanding, which first builds a model of reality and then tries to align it with the outside world—or vice versa, align the outside world with the model.

There's an apparent problem with this driving force toward maximum alignment, sometimes termed "the dark room problem."[1] If all that the cortex wants is internal coherence, you would think that the easiest way to achieve that would be to find a dark corner in a dark room: cut off all sensory input and nothing needs explanation or modification. And yet for some reason, we all leave our dark rooms and get up, do things, and seek out new experiences.

Clearly, the mechanism is incomplete: there must be something that pushes the cortex out of the dark room of nonexperience and into the world of novelty, surprises, goals, and achievements. As a matter of fact, there is another module of the brain whose entire essence is to orchestrate precisely such a push. It is called the reward system, and the main tool it uses to guide our decisions and motivations is a special neurotransmitter called dopamine, a tool both wonderfully clever and frightfully diabolical. Dopamine is what keeps us moving forward.

To understand what that means, it's helpful to look at what happens when you have no dopamine. A mysterious disease called encephalitis lethargica, which swept across the world from 1915 to 1926, presents a terrifying case study. It was most likely a complication of a common throat infection, which, in a small fraction of patients, caused their own immune system to attack the brain—but at the time, no one had any inkling of what was going on.

One of those patients was a young, wealthy New York socialite, later known by the pseudonym Rose R., who, in 1926, went to sleep and had a nightmare. In her dream, she was locked in an impregnable castle—or, rather, she *was* the castle. Rose woke up, but the dream lingered. She stared into space, saw herself in the mirror, but could not force herself to move. Her mind felt stuck in her own head, trapped in never-ending, remorselessly repeating chains of thoughts. The tune from Verdi's "Povero Rigoletto" looped in her mind endlessly, torturously, like a broken record. Rose's family tried to snap her out of the paralysis, but she remained motionless, her eyes open and a blank expression frozen on her face.

The nightmare went on, uninterrupted, for forty-three years.

Obviously, we wouldn't have known anything about the content of Rose's nightmares if she had not woken up eventually and told the story to the young New York neurologist Oliver Sacks.[2]

Rose was sixty-one years old by the time she was seen by Sacks, but she looked thirty years younger, as if cryogenically preserved, her face still and smooth, with no wrinkles. She would sit without moving for hours; occasionally she would twitch or utter a word or a phrase.

At that time, Sacks was a resident at the Mt. Carmel Hospital in the Bronx, which by 1969 housed about eighty encephalitis lethargica patients. Sacks noticed that some of their symptoms resembled an extreme version of another disease, Parkinson's. He decided to try a drug called L-DOPA, a promising new treatment for Parkinson's disease, on his encephalitis lethargica patients. If encephalitis lethargica was indeed a form of super-Parkinson's, Sacks reasoned, maybe L-DOPA could help them too. He turned out to be right. Within days of starting the treatment, patients awoke, rose to their feet, and began walking around, striking up conversations with stupefied hospital staff. Among them was Rose R., who woke up joyful, active, and fully conscious.

To Sacks's shock, the awakening was short-lived. For Rose, it lasted about a month. Some patients held out longer, but eventually their condition inevitably deteriorated. They developed tics and grew progressively paranoid. Over time, it became difficult for them to walk, then to move, and eventually they sunk into the same torpor as before, but this time an even higher dose of L-DOPA was not enough to bring them back. It was not until 1979, another ten years, that Rose choked on a piece of food, and her nightmare ended.

For reasons that remain unclear, encephalitis lethargica damaged the brain in one specific location: the substantia nigra, "black substance," a small cluster of neurons located deep within the brain, in a region known as the midbrain. The same region is affected in Parkinson's disease, so Sacks's intuition about the similarity between the diseases was incredibly insightful.

Substantia nigra is one of the few places in the brain that makes dopamine. Together with another adjacent cluster, known as the ventral tegmental area, the substantia nigra forms a dopamine-producing hub, where it is distributed to the rest of the brain—including the cerebral cortex. L-DOPA, the drug that Oliver Sacks used to temporarily bring Rose R. back to life, is a precursor of dopamine. It is used to compensate for the loss of neurons in the substantia nigra: by giving more of the precursor to the surviving neurons, some degree of natural dopamine production can be restored.

Basically, encephalitis lethargica shows what happens when the brain runs out of dopamine: it stalls.

This does not result in a complete coma. Patients were still able to chew food if it was placed in their mouths, for example. Some of them could respond to simple questions. Some would catch a ball thrown at them. They just didn't initiate any of this on their own, as if they had absolutely no motivation to perform even the simplest actions.*

These symptoms are consistent with what we know about other animals. Dopamine-depleted mice, in which dopamine production is genetically ablated, can also chew and swallow, but do so only if you place food directly into their mouths; they can respond to startling stimuli and hold tightly to your finger if you hold them up, but when placed in an arena in which a normal mouse immediately starts exploring, they just stand—eerily, without motion, not even flicking their tails.[3]

---

*Most terrifyingly, many patients appeared to form cohesive thoughts and experience emotions, as if they were locked inside a body that refused to move.

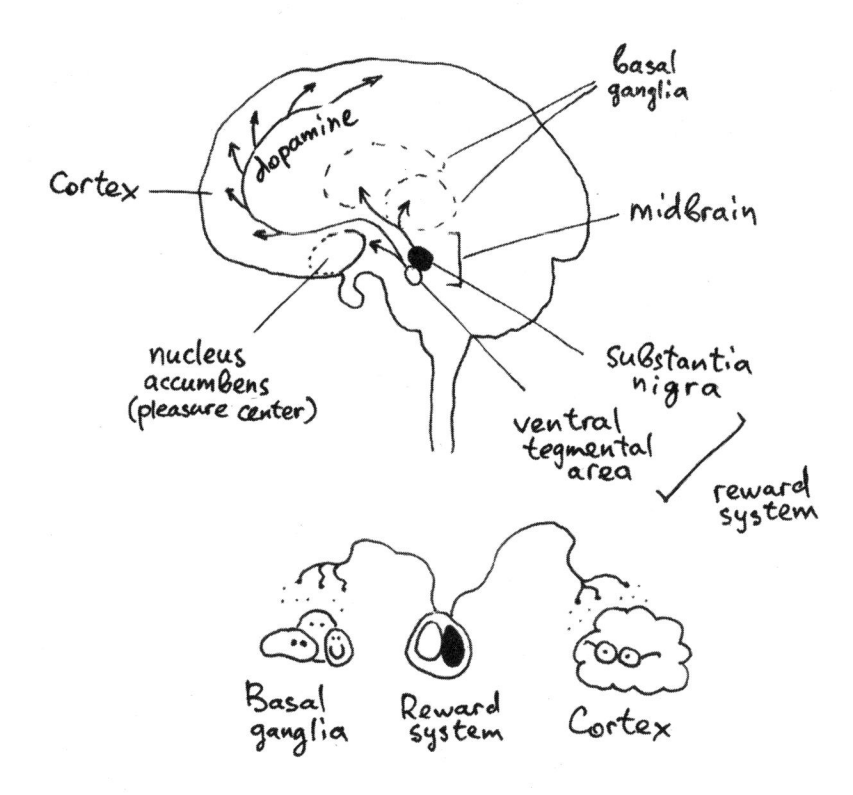

[Diagram: A hand-drawn illustration of a brain cross-section with labeled parts: Cortex, dopamine, basal ganglia, midbrain, substantia nigra, ventral tegmental area, reward system, nucleus accumbens (pleasure center). Below are three small sketches labeled Basal ganglia, Reward system, and Cortex.]

Removing dopamine from the brain doesn't simply paralyze it. Instead, it puts it in the dark room—a state of nonaction and nonexperience in which it does not feel compelled to do anything at all. That's exactly what you would expect from the cortex alone, if you accept that all it wants is to align reality and expectation. The moment you remove dopamine from the equation, it successfully achieves that by doing nothing.

So anything we do on top of basic reflexes, such as chewing the food when it is placed in our mouth, is motivated by dopamine. We would have all ended up in the dark room—the same torpor that paralyzed Rose R.—had it not been for the constant infusions of this chemical into our brains. Instead, we cannot wait to spend every waking moment of our lives in constant action. This is all because of dopamine.

So it must be dopamine's fault, then, that we spend every day battling with ourselves and always want to do the wrong things. If it's there to motivate us, why is it doing such a bad job?

To answer this question, we first need to understand what precisely dopamine does.

What Dopamine Means

There are a few ways to think about the essence of dopamine. The most basic way to understand it is "pleasure chemical." That explanation is helpful as a first pass, but it is wrong.

This idea works as follows: we do things, and when we succeed at them, we get a jolt of dopamine, which we experience as pleasure. We want more of the pleasure, so we continue to do the successful thing. As we achieve more successes, we find more and more sources of dopamine, and so our life gets broken down into various possibilities of dopamine acquisition, which become more and more sophisticated as we learn about the world. If we stop doing the successful thing, dopamine is withdrawn, and we desperately seek a way to replace it by doing some other successful thing.

The problem with this explanation is that dopamine doesn't actually cause pleasure. If you have a friend who takes Adderall (a drug used to treat ADHD that acts by squeezing out available dopamine from dopamine-producing neurons), they might tell you that the pills make them more focused, more productive, and put them "in the zone," but they don't produce euphoria.[4] Studies in rats say the same thing: an injection of amphetamine (the same type of drug as Adderall) makes them work harder for the rewards but doesn't increase their enjoyment, based on facial expressions and paw motions associated with positive and negative reactions.[5]

A similar but slightly more sophisticated take is that dopamine is a "do more of that" chemical. It's not about pleasure—it's about memory. It helps the brain remember which actions led to the successes.

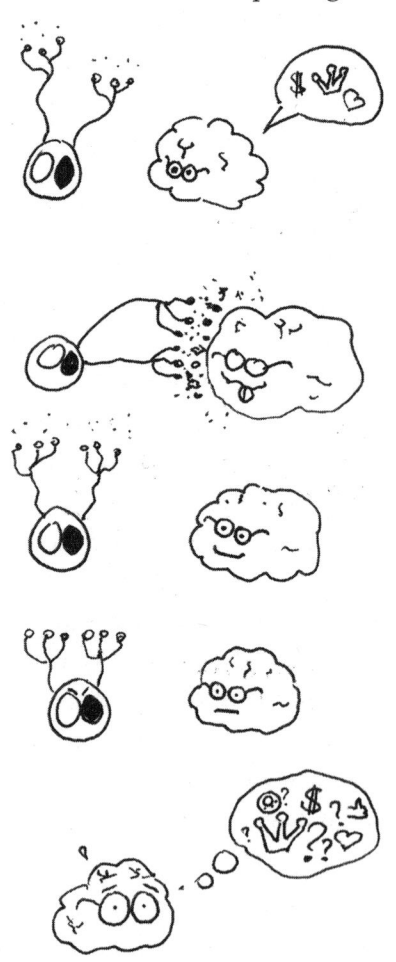

Dopamine does, in fact, boost brain plasticity, and so it enhances memory. The bodies of neurons that manufacture dopamine are located deep in the brain, but their dopamine-releasing endings can be found throughout the brain, carrying dopamine signals into distant regions, like a broadcast. This dopamine broadcast works in parallel with other neurotransmitters, such as glutamate, the standard signal that neurons use to pass excitation to each other. If a neuron somewhere in the brain receives a typical glutamate signal from another neuron, but at the same time also receives a dopamine broadcast, the glutamate-based connection grows in strength, as if dopamine is telling the brain: "in the future, do more of what you just did." So wherever dopamine is released, memories are stored better.

The clearest example of this "do more of that" role of dopamine is in skill formation, which occurs in a brain region called basal ganglia.

When someone is learning how to dance, they start by making each motion consciously. This means using the cerebral cortex to control the dance, motion by motion. As the cortex sends signals to the muscles—the leg bends this way and the hands that way—it also sends copies of those signals to the basal ganglia. If a combination of movements happens to land particularly well, a burst of dopamine is released by the midbrain, and this dopamine also goes into the basal ganglia. All neurons in the basal ganglia sense it, but it only affects those neurons that have just been activated—the ones that, an instant ago, received a signal from the cortex that led to the successful combination of movements. And so those successful neurons become "stronger" via dopamine modulation. In this way, through practice, dopamine selects successful dance motions and preserves them as a set, a unified combination that can be triggered all at once, directly from the basal ganglia, without the cortex having to think about every move. A skilled dancer then needs only to initiate this combination by thinking about the context—a particular moment in the song—and the sequence then "unpacks" itself, without conscious control. We call this "muscle memory"—in fact, it is basal ganglia memory, stored using dopamine signals that gradually optimize successful combinations of movements.*

---

*We've all been in situations when we accidentally switched from this optimized, basal ganglia memory back to the original, conscious cortical control of every motion, and immediately lost a skill. This is known as the "centipede effect"—based on a short poem in which a toad asks a centipede which of its legs moves after which, and the centipede immediately forgets how to walk.

The "do more of that" logic extends to other brain areas that receive dopamine, including the cerebral cortex.[6] Dopamine is released after something successful has been achieved; it strengthens the neurons and the connections between them that led to the success; we return to those neurons and those connections again and again. In the cortex, this might mean returning not just to neurons that execute an action, but to neurons that *think* about it—and so "do more of that" applies to thoughts, too, if we find them successful.[7] If you have an insight that suddenly illuminated a problem, you will get a jolt of dopamine, and the neurons that were involved in that insight would solidify their connections.[8] Next time, the insight will come more naturally. If a line in a song strikes an emotional chord, you will get a jolt of dopamine[9] and wake up the next morning to an earworm. In the same way as dopamine selects successful combinations of motions in the basal ganglia, it selects successful combinations of thoughts in the cortex. So if we do something dopamine inducing, we start thinking about it more. In fact, this is what our minds are constantly doing if we let them loose—searching for hidden dopamine in thoughts about the past or about the future. The more we think about something, the more likely we are to think about it again.

So, based on this explanation, dopamine helps us select the best actions and thoughts for achieving particular goals—do more of *that*, it tells the rest of the brain when a goal is achieved.

Except there is a twist: success doesn't always result in dopamine. Actually, what causes a burst of dopamine is not just any success, but *unexpected* success.

This changes the "do more of that" logic quite a bit. Here's how it works, for example, in a rat. Dopamine neurons are constantly releasing dopamine at low levels—like a gentle hum of a radio broadcast. Let's say you put the rat in a cage, flash a lightbulb, and then deliver a reward—a drop of sugar water. The first time you do it, the rat does not know what the lightbulb means, and there's no surge of dopamine the moment you flash it. But then the rat gets a reward—unexpectedly. This is when the dopamine surge happens—the volume of the broadcast suddenly increases, then returns to its normal volume. Next, you repeat the procedure many times. Gradually, the rat learns that the lightbulb precedes the reward. As it does, it starts reacting to the lightbulb—once it's on, the rat knows that the sugar water is coming, and dopamine surges. Here's the kicker: as more dopamine is released in response to the lightbulb, less gets released in response to the actual reward: sugar water. Over time, the only jolt of dopamine the rat gets is from seeing the cue. Once it actually gets the reward—achieves the success—dopamine stays steady, as if nothing happened. If you flash the lightbulb without delivering the reward—now expected—then the dopamine signal goes below the baseline: the radio hum momentarily turns into silence.

So dopamine release most closely aligns not with the actual reward delivery, but with the surprise: the more unexpected the success, the more dopamine. When the reward is delivered for the first time, it is at its most unexpected, and dopamine release is the strongest. As the rat learns, the lightbulb is what becomes "unexpected"—the rat doesn't know when it will flash—but once it flashes, the reward is expected. If something is expected but does not arrive—a violated expectation—dopamine dips below the baseline. So based on this, dopamine is a "better than expected" chemical, and its depletion means "worse than expected."

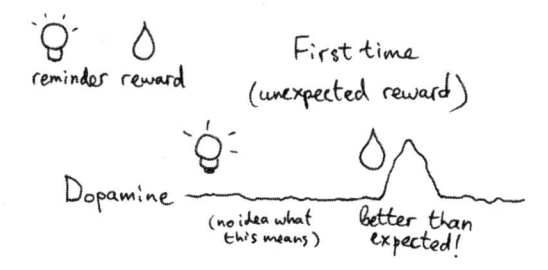

reminder 💡 💧 *reward*   First time (unexpected reward)

Dopamine ─────💡─────💧∧────
              (no idea what   better than
               this means)   expected!

After learning

💡∧   💧
better than (expected)
expected!

💡∧   ✗
better than
expected!   \/
              worse than
              expected!

The Dark Room **217**

This is a more nuanced explanation for what dopamine does than simply "do more of that" or "pleasure chemical." But it takes us back to the dark room problem.

Who decides what is expected and whether what is actually happening right now is better or worse than that? The cerebral cortex does.[10] No other brain region has enough information to piece together, for example, what money is—and money is a reliable source of dopamine in the human brain.[11] So it is the cortex that must tell the reward system about an unexpected success and in response receive dopamine. So, basically, the cortex stimulates itself through the intermediary of the midbrain, which distributes the dopamine.

But didn't we say that all the cortex wants is to minimize its own effort, to align reality and expectation? If there is something unexpected, then wouldn't the cortex want *less* of it, not *more*? Why, then, do we seek experiences that bring in new dopamine—travel to new places, read books, browse Wikipedia? Why is novelty something that we are drawn to, if all we want is to get rid of it? This is the dark room problem all over again—once you deny dopamine its essential "pleasurability," it becomes unclear why we seem to be driven toward things that produce it, or why we are driven to anything at all.

This is still an active area of research, and in my opinion, the precise relationship between the cerebral cortex and dopamine is one of the greatest unresolved questions in all of neuroscience.

Here's how I think of it, though I might be proven wrong in the future.

Dopamine is *not* what the cortex actually wants. Actually, what it wants is to *minimize* dopamine, just as it wants to minimize all of its activity. But, ironically, it gets dopamine any time it identifies a situation it deems unexpectedly successful—that's just how things are wired together! Rather than thinking of this dopamine jolt into the cortex as a positive, pleasurable signal, I think it makes more sense to think of it as an imperative signal: *figure this out*. This signal acts by accentuating the discrepancy that the cortex detects between expectation and reality—*if this thing is so good, how come I don't have it all the time?* That forces the cortex to do what it always does—find a way to eliminate the difference by either changing the expectation or changing reality. I would guess that dopamine must shift the balance of forces toward changing reality, compelling us to act rather than accept the state of things as they stand. But as of this book's writing, I don't know of any research that definitively shows that it does that.

What about our desire for novelty? If new things bring in dopamine, and we want to get rid of dopamine, why do we do new things?

Maybe what we actually desire is not the novelty itself, but the process of turning novelty into expectation. Surprise without resolution is simply confusion, and confusion is not pleasurable. But if the surprise makes sense, it delights us. It appears that we like resolving uncertainty more than we hate uncertainty itself.

And maybe the inevitable result of this pleasure that we derive from turning novelty into expectation is the reason we can never be satisfied. We don't enjoy rewards per se—we enjoy the process of sliding down the ramp from surprise to nonsurprise. So we inevitably get to the bottom of that ramp and inevitably find ourselves wanting to get back on it, but the ramp is now gone, and we have to find a new surprise. As we learn more and more about the world, we gradually expand our range of expectations and look for novelty in progressively narrower, more nuanced domains. When you are a child, you are happy to simply hang out outside, finding motivation in playing with a stick or in the $5 you earn at the lemonade stand. As you get older, you need a lot more just to stay content with yourself. You require an endless scroll of tailor-made TV shows to stay entertained. The sums of money that motivate you gain several additional figures. It's like Bitcoin mining, which was easy when the cryptocurrency was first invented but requires massive data centers today. Same with dopamine: the more expectations you create, the harder it is to find new surprises.

So overall, dopamine doesn't tell the cortex "good job." Instead, any time something turns out to be better than expected, dopamine says, "get to work and make this the expectation." If it doesn't say this, as with encephalitis lethargica patients, the disagreement between reality and expectation is simply not strong enough to warrant any action.

So, all in all, the best way to think of dopamine is as a "figure this out" chemical. This explains the effects of both amphetamines and dopamine depletion on mice. It explains why Adderall can create "tunnel vision" in human patients. It explains why people with low levels of dopamine experience lack of motivation.[12]

It also explains our fascinating obsession with uncertainty.

This is not unique to humans. Classic studies on the subject were done on pigeons[13] but have since been replicated with other animals, too.[14] You give these pigeons a button to peck and a reward as a result. Then you start changing the number of pecks required per reward. The more pecks required—say, fifty or a hundred pecks per reward—the more fatigued the pigeons seem after completing the task and the more reluctant they are to resume pecking.

But make the number unpredictable, and the pigeons don't stop. They continue pecking and pecking and pecking obsessively, regardless of how many times they get the reward. What motivates them is not the reward per se, but rather, a *pattern yet to crack*.

It gets even better. Say you once again take some pigeons, put them in a cage, and install a button, but this time you simply deliver the reward at random times regardless of any pecking. Soon, a few of the pigeons start pecking the button. Eventually, *all of them do*. They all dig in, trying to figure out a pattern when there's no pattern to figure out—and so they make it up, gradually becoming convinced that they are causing the reward. This is known as autoshaping—the pigeon equivalent of dancing for rain.

All of this sounds almost painfully familiar. This is precisely why gambling and social media are so addictive: not just the monetary or social rewards, but their unpredictability. You never know which of your photos on Instagram will get a lot of likes or which of your TikToks will go viral. Casinos and social media networks amplify this unpredictability by delivering the rewards at random times—they are certainly well aware of B. F. Skinner's experiments on pigeons that I just described. Imagine how it would feel if all your "likes" arrived together, once a week, at a designated time. You would probably come to dread the day—it would hardly ever feel better than expected and mostly worse than expected.

The essence of dopamine is not to make us happy and neither is it to direct us toward a particular thing, good or bad. It is to compel us to actively fit reality into a pattern and to make that pattern the expectation.

Wanting and Liking

What about pleasure? If not from dopamine, where does it come from, and why does it seem to align so closely with dopamine in most cases?

Part of the answer may be the very meaning we put into the word "pleasure." It could be that pleasure includes not just "liking," but also some "wanting"—the excitement due to an expectation of resolution that dopamine delivers.[15] So maybe dopamine is part of the sensation of pleasure, if not all of it.

Still, other brain chemicals, such as opioids (which include endorphins) and endocannabinoids, appear to be more in charge of pleasure—its most obvious, hedonic, "liking" aspect.[16] They are often released simultaneously with dopamine, painting the experience in a subjectively positive, rewarding light—maybe they should have been called the "reward system," not the dopamine, which, by itself, does not really reward us in any sense, but rather makes us work more.

As with dopamine, the effect of these "liking chemicals" also decreases with repetition—except usually not by reducing the release, but by reducing the quantity of endocannabinoid and opioid receptors on receiving neurons. In people addicted to opioids, for example, there is so much of the artificial "pleasure" signal periodically flooding the brain that the cells remove almost all the receptors for it, and so the person feels awful when the "pleasure" signal returns to normal.[17]

What I find most interesting—and least explored as of today—is how we come to enjoy complex things. Most of what we know about opioids and endocannabinoids is how they respond to primary, basic rewards—such as food or sex. It's clear that some of these responses are hardwired, or genetically preprogrammed. A lot less is known about how these chemicals contribute to liking more complicated things that must be learned—such as, for example, music. Do the same "pleasure chemicals" that we get from food and sex also make us like our favorite songs? To find out, researchers looked at how naloxone, an opioid blocker (also used to prevent overdoses of opiates), affected the perception of music. On the surface, people who received naloxone injections responded less strongly—their pupils dilated less and their skin conductance didn't change as much when they listened. But fascinatingly, their reports about how much they enjoyed music didn't change. Objectively, the lack of pleasure chemicals made music less effective, but subjectively, the pleasure remained the same.[18]

And so it appears that the conscious experience of pleasure does not always require any particular chemical—it is the pattern resolution itself that we call pleasure, and different brain chemicals just change how that pleasure manifests and what we do about it. Music is all about pattern resolution: both melodic and rhythmic patterns periodically intensify and then resolve. In music, we seek the same as when we travel to a new place: novelty to conquer, dopamine to eliminate—by listening over and over and over until every part of the song is so predictable that it becomes boring, and we move on to something else.

This is all very puzzling. On one hand, pleasure is easy to induce with simple molecules, such as morphine or cannabinoids. But on the other hand, no particular chemical seems to be absolutely critical for it. You can also induce pleasure by stimulating a particular brain part: for example, nucleus accumbens (a brain part most commonly labeled "pleasure center") or orbitofrontal cortex. But neither appear absolutely required.[19] In fact, our capacity to experience pleasure seems surprisingly robust. In one patient with extensive cortical damage due to encephalitis, multiple areas implicated in causing pleasure were affected, but the patient nevertheless retained normal, positive emotional responses to his doctors, friends, and family, saying things like, "I have a strong feeling of happiness, that we are here together working on these wonderful games and feeling happy together."[20]

To summarize, pleasure is a lot more elusive than motivation, but at least we can take hope in the fact that it's difficult to completely abolish.

Struggle as a Feature

It's starting to become clear why we seem so misaligned with our motivations no matter what we do.

Dopamine does not mark up the world into "good" and "bad." That would be easy: just do the "good" things, avoid "bad" things, and always stay motivated. Instead, dopamine marks an unexpected success—whatever we decide that means—and tells us "figure this out so you always have this success and are no longer surprised by it." *This* is the motivation, and it lasts until it becomes an expectation. So there's no trick for always being motivated: no matter what you decide you want, doing it for long enough will eventually demotivate you. You start a project that you are excited about, but then it turns into a chore, part of the routine. And so your brain drifts away from it toward other unresolved discrepancies between prediction and reality—are there, for example, video games that haven't been played yet? The only way to maintain motivation is to keep moving on to new things.

This is a brilliant system, as far as its evolutionary value is concerned. Imagine two animals, one of whom is perfectly content with what it has and the other who easily gets bored and constantly looks for more. Which animal is more likely to survive in the long run? The one that goes beyond what it has—because even if you have everything now, nothing ever lasts forever, and you will be better off if you prepare for this in advance. Dopamine is a bet on inevitable future change. Evolution favors the restless, the unsatisfied, the novelty cravers tormented by visions of more, because that keeps them from settling into place and, in the end, ensures their greater success.

As for peace of mind—well, you can live without that.

The reward system is agnostic to what we like—it enacts a desire for food or water in the same way that it enacts a desire for money. But in all cases, it eventually turns everything new and exciting into a boring expectation that we depend on, but take it away, and we feel worse than we were before we knew about it. A newborn baby does not yet know that the mother's breast provides nourishment, so it can't desire it. All it can do is suck when the breast touches its mouth. But the moment the baby realizes the surprising connection between the sucking and the nourishment, dopamine creates a desire. Now, not having the breast at all times is a burden that the baby has to carry. And so it continues for the rest of our lives, as we find more and more complicated ways to achieve success and, in doing so, grow frustrated with more and more things that dopamine forces us to incorporate into our models of the world.

Dopamine neurons convert good things into dopamine

There was once a young nobleman in India who arrived at very similar conclusions without any knowledge of neuroscience. His name was Siddhartha Gautama, and eventually he would be known by his enlightened name—the Buddha.

Siddhartha saw that happiness doesn't seem to depend on what one has. Satisfying desire only leads to more desire. "Humans are prone to suffering," concluded the thoughtful Gautama, who left his home on a quest to find a solution—this is how Buddhism begins, according to tradition.

It's hard to argue with the premise. In modern terms, Buddha taught that human nature always strives for what it does not have and always suffers as a result. If you satisfy one desire, you develop another, bigger one. If you solve one problem, you will have to solve ten more. If you fix one error, all errors call upon you to be fixed, and since this is impossible, you inevitably suffer. So the only way to not suffer is to not resist anything and not desire anything, concentrating the mind on the present moment and accepting it for what it is. Buddha's "enlightenment" was not a moment of rapture, but a gradual process of mastering meditation and tuning out unnecessary desires. "Nirvana" actually means "blowing out," as in blowing out the candles to see a dim light.

It's true: the very process of wanting something gradually brings about a dependence. First you feel good when you have it, then you feel bad when you don't. Over time, unexpected pleasures inevitably turn into expected needs. It is fun to beat a rival team at football, but winning every time gets boring, and losing becomes an insult. To experience the joy of victory afresh, the bar needs to be set higher—the regional tournament. A loss would mean embarrassment and a happy return to the local rivalry. But victory, although rewarding, would push the bar even higher, another step on the infinite dopamine ladder of mounting desires and their satisfaction. Eventually, all paths lead either to suffering or to the escalation of desire.

So it's just like the Buddha said: pleasure begets desire, and desire begets suffering. The goal of the reward system is not to make us happy, but quite the opposite: to keep us unsatisfied.

You could say that the cortex knows everything but doesn't want anything, whereas the reward system wants but doesn't know. In order to know and to want, they have to work together. Dopamine is like a carrot hanging in front of a donkey—it's still the donkey who decides the direction, but without the carrot, it doesn't move anywhere.

This dichotomy actually precedes the very existence of a brain. It goes way back to the beginning of the eukaryotic domain, in particular to something discussed in the first part of the book: the advent of multicellularity.

What we said was that the transition to a multicellular organism gave us the gift of individuality. It allowed us to be partially independent of our genes. All the instructions for all our trillions of cells must fit into a single egg and a single sperm—there is simply no way to preprogram every single element of such a massively complex organism, and so it must, to some extent, fend for itself. The more complicated it gets, the more freedom it acquires on top of the genetic instructions. Eventually it grows a brain and starts learning from the environment, not just the genes. That's where this essence—individuality—really starts to flourish.

And yet, genes keep the body on a leash. If the body refuses to eat or reproduce, it will simply die out. So evolution preserves only those bodies that retain some degree of obedience. No matter how much we learn for ourselves, there must be some instinct that reminds us about the fundamentals: don't die, eat food, have sex.

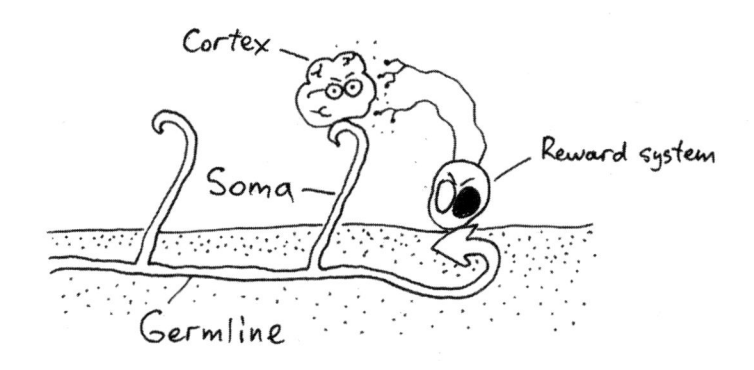

It is as if we had two separate sources for our desires and motivations. One is our cortex, the thinking machine, the essence of which is to learn from the environment, understand its patterns, and act based on this understanding. All it wants is predictability and peace—and the process of getting there is what we call liking. The second one is the reward system, the wanting machine, the essence of which is to push us out of this comfort zone and make us look for more. It has long abandoned any attempt to understand what it makes us want—that's on the cortex—but it will continue to push us to want it, because that, in the long run, helps us survive and carry the germ line forward. The reward system is like the germ line's representative within the soma, executing genetic will of our ancestors through waves of desire and suffering that it spreads throughout the brain.

The cortex on its own would probably quite enjoy the dark room—a state of nonactivity and nonexperience. But pestered by the reward system—the past, the habits, and the memories teasing it with unpredictable reward—the cortex is forced to seek it out, to calculate and plan, to look out for danger, and to plot the next move. The dark irony of our existence is that we seek two different things at the same time. On one hand, we want peace and predictability. On the other hand, we want surprises and pleasures. The only path to long-term happiness is to somehow reconcile the two.

The Buddhist Economy

So how do we save ourselves from suffering? Is there any way to break the cycle of desire and dependence? Siddhartha's method was rooted in the control of attention—meditation. Humans can direct their attention where they decide. Without such control, we burrow through our own minds trying to find satisfaction either in the past or in the future and never feel satisfied. Reminding yourself to remain in the present moment means that you don't savor the victories but also don't mind the defeats; there are less pleasures but also less dependencies.

The quieter the background hum of stimulation, the more stimulating the everyday activities and occurrences become: talking to friends, watching a squirrel outside your window, even breathing. It is, however, exceedingly difficult to rid yourself of the habit of thinking about the future or the past.

Dopamine distracts Buddha from meditating

Unlearning all the joyful and stimulating experience requires a reprograming of the cortical model. This is what Buddhist practice is fundamentally about.

If that seems like too much of a commitment, you can reframe the same ideas in economic terms. Think of dopamine as a finite resource—like a salary. You only get so much of it per day, per week, per month. You need it to motivate yourself. But—and this is the key trick that our brains are doing to us—you also spend it any time something really good happens. So you have to balance the budget between motivating yourself and rewarding yourself.

Spend all the dopamine on social media, algorithmically optimized to squeeze it out of your neurons, and you'll have none left to motivate the less stimulating activities like reading books. Save up for a big party and use it all in one go to have the night of your life, but be prepared for a few dreary days. Avoid unnecessary spending—and always have some pocket change for extra motivation.

Interestingly, this applies not only to actions, but also to thoughts. You spend your dopamine even when you *think* about a good thing. The more a person wants something, the more dopamine is spent on playing the thought on repeat. The thought becomes baked into the brain—a firm expectation. When people spend years dreaming of a specific goal, upon reaching that goal, they often feel numb and occasionally deeply disappointed. This is what F. Scott Fitzgerald's *The Great Gatsby* is about: sometimes it is best for dreams to remain dreams.

A happy life is a compromise between the free-spirited cortex and the dictatorial reward system. Not everyone needs to rid themselves of all desire. All we need to do is balance the budget. In the long run, it is not important where we derive our satisfaction—how we handle it makes a lot more difference. Our behavior pushes us toward fixing problems and achieving goals, but that road does not lead anywhere. No goal-directed act can bring long-term satisfaction, because reaching the goal always creates a new goal—we react to changes, not to specific states.

All of this means that happiness can be found only in the process, not in the result. The goal of life is not to solve a riddle, but to suspend yourself in the question—like a Zen koan.

Chapter 12.
In the Beginning Was the Word

> **Everything passes, but not everything is forgotten.**
> IVAN BUNIN

In Jewish folklore, a clay man—a golem—comes alive when you place God's name, inscribed on a piece of paper, under his tongue.

In the first two parts of this book, we traced the essence of our own species from the conception of life on Earth almost to the present day—*we* as living organisms, as eukaryotes, as animals, as vertebrates, as primates. Right at the end of this journey, we took a detour to learn about the essences that make up our inner, conscious worlds: *we* as abstraction, as memory, as understanding, as wanting.

It is now time to add the final essence into this Russian doll of "we's" that forms each of our individual "I's": language.

Like God's name placed under the golem's tongue, language is what truly brings the *matryoshka* to life—not life in the biological sense, but life in the human sense, the life each of us experiences in first person.

Most of us consume tens of thousands of words per day—news, conversations, emails, advertisements, books, social media, TV shows, music. Most modern jobs, at least those in developed countries, involve some form of word processing: office work, healthcare, retail, information technology. We dive into words the moment we wake up and drift through them until we go to sleep. We occupy ourselves with concepts that simply do not exist without words. Presidential elections. Climate change. Food and drug administration.

To be sure, other living creatures use a variety of "languages" to communicate information. Trees exchange chemical signals alerting each other of a parasitic fungus attack.[1] Lobsters identify each other using custom cocktails of pheromones called "signature mixtures."[2] Birds sing songs to one another. Many ocean inhabitants emit some form of light to either attract a mate or deter a predator.[3] Dolphins come up with unique whistles that they use like names.[4] Vervet monkeys have different "words" that alert group members of different threats.[5]

And yet all these protolanguages are different from human language in one critical aspect: no matter how complicated they are, they communicate only a limited number of ideas. Maybe a plant can send another plant a dozen different chemicals meaning different things. Maybe—this would require some imagination—the number could reach a hundred. It is even possible that some of these chemicals mean different things in different contexts: perhaps there is a large variety of combinations that the plant can respond to in different ways.

But that would probably be it. Every chemical signal would have to be received by a dedicated receptor, with a corresponding gene in the plant genome; every combination would have to be honed by evolution and optimized to produce a particular response.

As far as we can tell, no plant can compose a poem, perform a scientific study using chemicals, or teach another plant something that it doesn't know and doesn't expect. Vervet monkeys might have different words for *snake* and *eagle*, but they don't combine them to invent the word *dragon*.

This is possible only with human language.

What makes human language unique is that it is infinitely generative: through grammar and syntax, you can create new meanings using existing words without any limitation whatsoever.

And that is what awakes the golem.

What gives human language its unique generative power? According to one influential school of thought, the distinguishing property of human language is recursion: the ability to embed structures within similar structures indefinitely.

Consider the nursery rhyme "The House That Jack Built." It begins like this:
This is the house that Jack built.

This description incorporates several strands of information. It invokes an object (*house*), clarifies that this object is a specific one, rather than one of many (*the house* rather than *a house*), describes its origin (*that Jack built*), and attributes it to an individual.

Next, all those strands of information become a package, embeddable into another structure:
This is the malt
That lay in the house that Jack built.

This is a new description of a different object. It now incorporates the description of the house as one piece of information among several—another being that it is a *malt* and that it *lay* inside the house.

This continues for another nine iterations, resulting in:
This is the farmer sowing his corn,
That kept the cock that crow'd in the morn,
That waked the priest all shaven and shorn,
That married the man all tatter'd and torn,
That kissed the maiden all forlorn,
That milk'd the cow with the crumpled horn,
That tossed the dog
That worried the cat,
That killed the rat,
That ate the malt
That lay in the house that Jack built.

If you simply said, "this is the farmer," you would have technically told the same story as this entire verse—identified the same person. But through recursion, the story can be inflated to incorporate a great deal of knowledge about the farmer. You can imagine it as a hierarchical tree in which different strands of meaning (*house, Jack, built*) merge together into new levels of meaning (*house that Jack built*), which then merge with other strands to form meanings of progressively higher order (*malt that lay in the house that Jack built*). The story is embedded in the nodes of the tree.

Even individual words and phrases, at close inspection, turn out to be complicated, hierarchical trees of meaning. Think of what it takes to understand a simple phrase like *talented rock musician*. You must understand the idea of *talent* (a natural endowment), the idea of a *musician* (a professional music performer), and have at least a cursory awareness of the musical genre also known as rock and roll. This is complicated enough. But it keeps going: to understand what "natural endowment" is, one must further understand the meaning of *natural* (as in, innate, not learned) and *endowment* (special ability to achieve success). This is contingent on even deeper understanding of ideas like ability and success—and

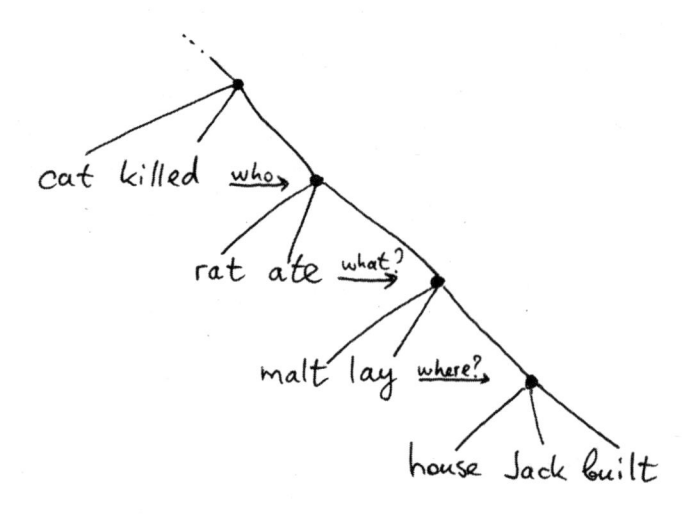

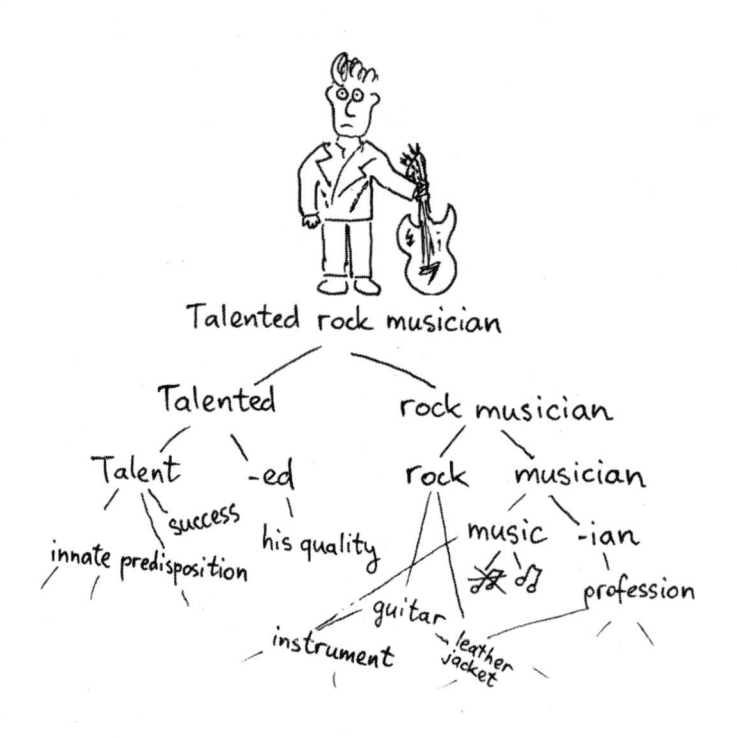

232 One Hand Clapping

so on. By saying *talented rock musician*, we package this vast hierarchy of meaning into a highly compressed, concentrated form.

This is simply not possible without language—if you don't have the words, you can't compress information in your brain to such an extent. So this recursive, hierarchical nature of language does not just allow us to *communicate* complicated things. It allows us to *think* them in the first place.

At least, such has always been the central tenet of Chomskyan linguistics—the predominant school of thought in the science of language of the past half a century.

The Special Sauce

According to Noam Chomsky, a towering figure in linguistics since the 1960s, language is not the ability to say and understand words, but the ability to construct an unlimited number of meanings out of them. Chomsky often quotes Wilhelm von Humboldt, the Prussian philologist of the century prior: language makes "infinite use of finite means."[6]

Chomsky insists that language is first and foremost a special type of cognitive process, a unique ability of the human brain. We may absorb words from the environment, but language—what we do with these words—is innate, installed in us by evolution. Speech and communication, in Chomsky's view, are secondary—almost a side effect, a result of language "externalization," when cognition breaks out of the confines of the brain.

One of Chomsky's main points is that virtually all children easily learn a language without any complicated instruction, simply by listening to other people talk. If you think about it, what kids actually hear is a hideously unpredictable jumble of sounds that's not even broken into individual words (in casual speech, we do not separate words as we do on paper, and so the very existence of words is far from obvious). Out of this sonic chaos, the developing brain somehow picks out the meanings of thousands of words and within a few years, almost without feedback, learns to create its own new meanings out of them.

Another point is that all languages share a set of structural similarities, a universal grammar, the main element of this grammar being recursion. According to Chomsky, it is this embedding of elements within elements—*the malt that lay in the house that Jack built*—that underpins every natural language as well as human cognition. You may think of recursion as obvious, but that's simply because all the languages you might speak have it. Computer languages, for example, are not necessarily recursive.

What both of Chomsky's arguments suggest is that language, at its core, is innate. We must all possess some sort of recursive "language organ," which, in kids, is tuned to the patterns of surrounding speech and rapidly populates an already existing mental scaffold with words from the environment. This mental scaffold is universal among humans and, as a result, all languages follow the same universal grammar, even if they use different words.

This hypothetical "language organ"—and its special ability, recursion—are at the center of the Chomskyan school of thought.

If we agree with it, it might seem that we are finally zeroing in on the nature of human uniqueness. What makes us special is all those infinite meanings we generate, and that is possible thanks to recursive language. This must be it, then: recursion is the special sauce of the human species! Did Noam Chomsky solve the mystery that eluded scientists since the nineteenth century, when Charles Darwin made us just one ape among many?

Not so fast, say Chomsky's opponents.

Hidden deep in the Amazon jungle is one mysterious tribe whose unique language challenges the Chomskyan dogma of recursion as the centerpiece of humanness. This tribe, Pirahã, takes us right back to the drawing board.

Crooked Head

Many linguists and anthropologists consider the Pirahã some of the world's most unusual people, but you wouldn't know that based on appearance. Typical Pirahã housing consists of a small bench and a simple canopy made of leaves or canvas. A few of those huts on the riverbank make a village. The Pirahã wear plain clothing such as T-shirts, shorts, and skirts. There is nothing special about their physical features. To strangers, the Pirahã appear thoroughly unremarkable. But get to know them a little better, tells US linguist Daniel Everett, and you will find that they inhabit a very different world.[7]

Everett's name is associated with almost everything known to the rest of the world about the Pirahã. He lived in a Pirahã village, on and off, for thirty years, first as a missionary, then as a scientist.

According to Everett, Pirahã don't plan for the future and don't preserve food in any organized way. There's no centralized distinction between night and day—sleep is considered bad for you. There are no strict rules pertaining to private property. The Pirahã are rarely interested in material goods. Children do not have permanent toys. They don't keep any art or personal possessions beyond their cotton clothes—virtually the only item from the outside world that they enthusiastically embrace. Pirahã men and women do form more or less permanent couples, but divorces are allowed and not frowned upon. There's no mythology. If you ask the Pirahã where everything came from, they reply, "It has always been like this." They seem almost invariably happy. They constantly smile and laugh. What they find most funny is all the outsiders, with all their endless questions, problems, and things. Pirahã have a special term describing all non-Pirahã, which translates roughly as "crooked head."

The most unusual trait of the Pirahã, though, is their language.[8]

The language has very few sounds, which gives it a repetitive, almost meditative feel, resembling at times birdsong and at times a Morse code stutter. Sometimes half a word is dropped. You can "speak" the Pirahã language by whistling if you want.

In terms of vocabulary, the language lacks key concepts that other cultures consider essential to communication: no "hello," "thank you," or "sorry"; no fixed terms for colors; no numbers. Many nouns are borrowed from neighboring tribes or from Portuguese.

But most notably—and this is the part that calls Chomskyan dogma into question—Pirahã appear to have no recursion.[9]

For example, when talking about the weather, a Pirahã might say a phrase that sounds something like this (there is obviously no writing): "*Aí, ka'aí ka'ai-o, abá-ti, piiboíi-so.*"[10] This literally translates as follows: "*Well, house in-house, I stay, it rains.*" So instead of using a complex sentence, "*I stay in the house when it rains,*" Pirahã speakers simply list the parts of their thought: *house, stay, rain*. Especially interesting is the construction of the verb *piiboíi-so*. The *-so* suffix indicates temporal distance, meaning that an action occurs either in the past or in the future, but not in the present: *it rains, but not right now*. Rather than explicitly indicating, through syntax, that rain is the condition and staying at the house is the consequence, the Pirahã assert instead that staying at home coincides with rain, in the past and in the future.

This grammatical simplicity is evident even when the Pirahã speak Portuguese. Some men have enough vocabulary to explain themselves, but they typically use Portuguese words in list-like Pirahã sentences, as in "Well, rain, drip-drip-drip, be at house, sit bench, well, sleep."

This, says Daniel Everett, is evidence that recursion and universal grammar are not so universal after all. Pirahã don't form embedded clauses, like *the malt that lay in the house that Jack built*. The fact that the rest of us do this, then, is not an essential part of being human, but simply one way to talk among many.

But if recursion is not an innate feature of our brains, then the whole Chomskyan notion of a "language organ" falls apart. If our minds must learn even the most basic features of language from scratch, then there's no special mode of cognition that is uniquely human, no predetermined mental scaffold. The brain is a blank slate.

So who is right? Is human language, in fact, an innate thinking pattern that gets externalized as speech (as Chomsky believes), or is it a cultural agreement that gets internalized to form distinct thinking patterns (as suggested by Everett and the Pirahã)?

Our best chance at reconciling the two schools of thought is found in the most surprising of places: a boarding school for the deaf in Nicaragua.

The Birth of Language

In 1977, Nicaragua opened its first boarding school for the deaf.

Prior to that point, the majority of deaf people in the country had been isolated from one another. Typically, everyone in a deaf person's family and community was able to hear and speak. As a result, most deaf people grew up without access to sign language, so they remained without a language at all. Usually, they worked out simple ways of communicating with their families by making up their own signs.[11]

At first a few dozen and later hundreds of deaf children from preschool to sixth grade started to come together at the newly established boarding school.

Initially, the school tried to teach the children lipreading in Spanish, reasoning that this would give them a better chance at integrating into Nicaraguan society. This proved to be impossibly difficult. But then the teachers observed something remarkable. The primitive, personal signs that each of the children had invented on their own to communicate with their families were being rapidly replaced with common gestures understood by all their student peers. An assortment of personal protolanguages was spontaneously crystallizing into a unified, communal language. It would become known as Nicaraguan Sign Language, or NSL, making history as the only natural language whose birth was documented by linguists from the very beginning.[12]

The new language started simple, but over time the number of signs increased and their meaning expanded. Soon, there were signs representing objects and signs representing actions—nouns and verbs. Multiple signs were arranged into sentences, sometimes with multiple levels of meaning. Abstract categories, such as time, and grammatical forms, such as perfect and imperfect tenses, started popping up. New students joining the school learned the latest iteration of the language without having to learn its previous versions. Within a few years, a primitive system of signs based on pointing and miming developed into a fully formed language, which today is used by thousands of Nicaraguans of different ages.

What exactly makes this case so unique? It's the fact that a few hundred children, all individually deprived of language, were suddenly plucked out of isolation and concentrated in one place. This typically does not happen. It is extremely rare for a child without a cognitive disability not to acquire any form of language, let alone to live among hundreds of children in a similar position for an extended period of time. The reason it had to be a sign language was only because if the kids weren't deaf, they would have never ended up in this situation.

The important part is not the sign language part. What NSL clearly illustrates is a profound truth: *a sufficient number of humans put together will invent a language.*

So on one hand, NSL seems to prove that Chomsky's "language organ" is alive and well. There's clearly something innate about our linguistic ability, given that the Nicaraguan children were able to manifest it after all the attempts to teach them an existing language had failed. The birth of NSL seems a lot more like externalization than internalization.

But on the other hand, this case vividly shows that our cognition depends on the specifics of the language that we internalize.

In one study, the Nicaraguan boarding school's graduates were shown the same sequence of cards with cartoons arranged into a plotline.

There are two brothers: a mean older brother and a wimpy younger brother. The bully brother grabs a toy train from the younger brother, puts it under the bed, and goes into another room. Meanwhile, the younger brother finds the train under the bed and, with a smirk, puts it into a box across the room. Then he runs away. The older brother returns, holding a toy rail pensively as he stands in the middle of the room.

Question: where will the older brother look for the toy train?

Participants were offered two resolutions to the story: the bully crawls under the bed where he left the train (this is, obviously, the correct answer), or he goes to the box across the room where the younger brother put it. Younger participants (i.e., more recent graduates) answered correctly 90 percent of the time. But the group only a few years older answered correctly only 10 percent of the time.[13]

How did they explain their choices? The younger cohort reasoned as most of us would: the older brother would search under the bed, because *he thinks the train is under the bed*.

But the older cohort reasoned differently: the older brother would search in the box, *because he wants to find the train*. When asked to describe what happened in a variety of similar situations, they almost always used signs describing desires—what the person wants to accomplish—and hardly ever signs describing mental states—what the person thinks or believes.

What is especially remarkable about NSL is how quickly it evolved. In only a few years, it developed from a primitive system of pointing and miming to a complex language with grammar and syntax. Each student attended the school for only four years and then left, carrying the current version of the language with them into their adult lives. So even after graduation, depending on the years of attendance, former students spoke slightly different versions of the same language: older ones spoke more primitive NSL, whereas younger ones used a more advanced version with more signs and more complexity.

It appears that at some time during the second half of the 1980s—the time period separating the two cohorts who attended the school—NSL acquired a sign for *think*. This upgraded the language—students could now talk about what someone else was thinking, rather than what they wanted, which was the only thing they could describe before. What the study shows, with unusual clarity, is that being unable to describe something also means being unable to grasp it. If you don't internalize the correct word, you have no access to an idea—an idea most of us consider very basic, logical, and independent of language.

This seems to point away from the Chomskyan notion of language as externalized cognition. Instead, it appears to align more closely with internalization: culture etching itself onto the blank slate of the brain, defining even the most elemental thinking patterns.

Pirahã are another example of how deeply language affects cognition. For instance, they don't have words for specific numbers—they have terms for only "few" and "many," which could mean totally different numbers in different situations. At the same time, they are hopelessly poor at the simplest arithmetic. If you place ten balloons in front of them and ask them to lay out the same number of batteries in perpendicular orientation, they fail every single time.[14]

This idea, language shaping cognition, is known more broadly as the Sapir-Whorf hypothesis, named after the 1930s researchers who first proposed it. Sapir and Whorf, who argued that human cognition is structured by the language we internalize, were thought to have been toppled by the great "externalizationist" Noam Chomsky. They were pronounced unscientific in approach[15] and later proven wrong about some of their claims regarding the Hopi language.[16] But today revised versions of the Sapir-Whorf hypothesis are popular once again.

Evidence from both NSL and the Pirahã language clearly supports the notion that our patterns of thinking are deeply impacted by the language we learn.

The key question is—just how deeply? Does the language we learn only teach us words, or does it also teach us what to do with them? This is what the Chomsky-Everett debate is about. One side says recursion is built in. The other says that even recursion has to be learned.

So does NSL have recursion? It appears that the answer is yes—but not much. To my knowledge, no linguist has described NSL sentences as multi-leveled as in "The House That Jack Built." But there does seem to be a way to embed a clause into another: when describing what someone did in third person, you represent their actions as if they were coming from the direction of your own body, shifting the frame of reference onto yourself and incorporating this first-person demonstration into the narrative about the other person.[17] What's especially interesting is that this shift in perspective became a feature of NSL only after several years of evolution—to the extent that older signers could not fully understand what the younger ones were saying when the shift came up in conversation.

Maybe, then, recursion is a natural feature of any human language, but it is not as categorical and dramatic a feature as Chomsky suggests. There can be more of it or less of it—and sometimes, perhaps, as with the Pirahã, the amount of recursion can be so low that it is almost undetectable.

Garlic Butter Bacon Cheeseburger

What we must do to resolve the Chomsky-Everett debate is to acknowledge that recursion in language is one specific case of how brains work in general.

To me, recursion seems like a familiar concept, and to you, the reader of this book, it probably does too. Except we have called it *abstraction*: the progression from specific to general, from small to big, from the underpinning to the overarching. Whatever you choose to call it, I think that linguistic recursion is just a specialized case of abstraction that forms the framework of our entire mind.

In this book, I wanted to first introduce the concept from outside of language. This is why we detoured into sea slug brains and talked about how *touch to the tail* and *touch to the head* combine into a more abstract essence, *dangerous touch to the body*. This is why I used mushroom foraging as an example of moving through the tree of abstractions in my own brain—it's intuitive, wordless, based on colors and shapes combining into distinct categories of "good" and "bad." We've said that our whole mental process is climbing up and down such inverted trees of abstraction.

By and large, though, we label abstractions using words, giving a name to each node of the tree, and those linguistic trees are the ones we spend our time climbing. Let's say I tell you about that amazing *corner bodega garlic butter bacon cheeseburger*. To understand what I'm telling you based on this sequence of six nouns, you must re-create, somewhere in your mind, the following tree.

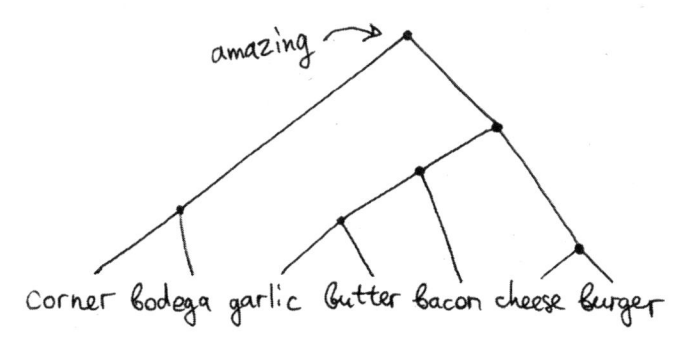

In other words, you have to grasp that *corner* is a description of the *bodega*, *garlic* modifies the *butter*, the resulting product modifies the *bacon*, which then modifies the item created by modifying a *burger* with *cheese*. Once you get it, we're on the same page.*

Our mental process largely consists of traveling up and down that language tree. This is the tree that you climb when you combine words like *house*, *Jack*, and *built* into a *house that Jack built*.

But the tree is bigger than language. We also climb it when we see lines on a page, which combine to form shapes, and shapes combine to form objects. We also climb it when frequencies of sound that we hear combine into notes, which combine into songs, which combine into styles of music. Language is only one segment of this branching tree, but one that is uniquely dense in the number of branching levels and interconnections among nodes.

So recursion—or abstraction, or whatever you prefer to call this progression from small to big—*is* an innate feature of our brain. And not just our brain—

---

*In Russian, you can famously construct a complete, fully conjugated, and occasionally useful sentence out of five verbs and nothing else: *decided send go buy drink*. You *can* do this in English (*help try making working disappear*), but it's a lot more strained.

all brains. Languages can utilize this ability of the brain and amplify it but not necessarily to the same extent. NSL started as a pointing system with no recursion but then began incorporating some recursive elements. Pirahã happened to incorporate comparatively little recursion. But if we think of recursion broadly, as the embedding of smaller meanings into bigger meanings, I would argue there is *some*: recall, for example, the term "piiboíi-so," which consists of the word "piiboíi" ("it rains") and the suffix "-so" ("not now," meaning either in the past or in the future). That's a new meaning created from a combination of two other meanings, a node on the mental tree that combines two branches. Pirahã language trees just don't have very many levels—they are "flatter" than most languages.

Escape Velocity

Unfortunately, this détente I negotiated between Everett and Chomsky—if they would accept it—means that we have lost Chomsky's special sauce for the human species. If recursion is not a recent evolutionary invention but simply something we humans are very good at, the question of what makes us special seems once again unresolved.

But I think the Nicaraguan school for the deaf can help us there, too.

What NSL shows is that children not only have an innate capacity to understand and produce language, but they also have an innate capacity to *pass it on*. Without this, the language would have never caught on, evolved, and continued to exist after the first generation of students left the school.

There have been cases in which other primates have been taught some variations of human language, including sign language. The University of Tennessee reportedly housed an orangutan so apt with sign language that at one point it asked to go for a ride in a car, brought along money it earned by keeping its room clean, and gave the driver directions to the local Dairy Queen.[18]

There are even documented cases of one ape learning such signs from another. A chimp named Washoe had a vocabulary of 350 signs and passed on a total of 55 to her adopted son, Loulis. He used the signs to say things like, "Hurry, come tickle" and "Give me that hose," referring to a water hose that the chimps liked to play with.[19]

But that was it. Loulis did not pass on the trick to others; the signs did not take off; the language was forgotten.

If we look at this case and then compare it to the Nicaraguan school for the deaf, I think we can glimpse the origins of humankind.

Humans appeared when language reached escape velocity.

If you shoot a cannonball out of a cannon, it eventually falls on the ground. The stronger the blast, the faster it will fly, and the further it will go before landing. But there is a point, a critical velocity called the escape velocity—for Earth, it's 11.2 kilometers per second—when the cannonball flies so fast that it escapes the planet's gravitational pull. So, at 11 kilometers per second, the cannonball flies pretty far but still falls on the ground, but at 12 kilometers per second, it does not fall at all. It starts circling the planet—potentially forever. That's how satellites are launched into orbit.

This is what happened with our species. Like all primates, our ancestors were primarily shaped by living in groups—this is probably why we have such a large cerebral cortex, as we discussed in chapter 8. There might have been relatively advanced forms of communication long before Homo sapiens. At first, they were probably created—and soon forgotten—by individual people and small groups.

But at some point, language *took off*—in a way that had never happened before, in a way that, eons later, another language would once again take off among Nicaraguan children. Not just communication in general, but a specific system of symbols spread across many people. Today we might say it "went viral."

Maybe that's the best way to think of it: language as a cognitive virus. It infects almost every human in their early childhood, using us to pass itself on to others. At the same time, it installs into our mind an entire operating system, which allows us to think at a level of abstraction no one in the world had ever contemplated. It's a symbiosis. We benefit from extraordinary cognition; the language virus benefits from getting passed on.

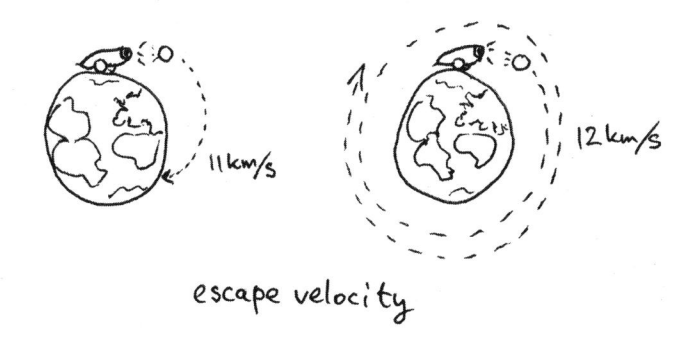

escape velocity

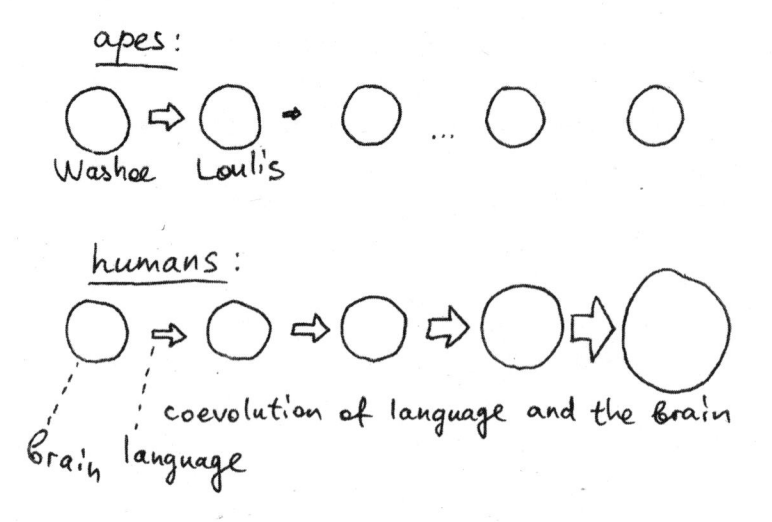

[Handwritten diagram:

apes: ○ ⇒ ○ → ○ ... ○   ○
        Washoe Lou‌lis

humans: ○ ⇒ ○ ⇒ ○ ⇒ ○ ⇒ ○
         brain language coevolution of language and the brain*]

The first form of language that went viral—reached escape velocity—must have been relatively simple, like the original version of NSL. But over a long time—probably millions of years—it evolved, owing to its ability to get passed on. Just like genes evolve because they could get copied, languages evolve because they get passed from person to person.

More complex languages means more complicated cognition—and so, increasingly, language did more than simply help people exchange information; it also helped them think and invent new things.

But as language changed and became more complicated generation after generation, it must have also changed the brain around itself, like particularly successful software forcing engineers to redesign the hardware. As language evolved, so did the human brain.

This is why today, language is so universal—we are adapted to this cognitive virus on a physical level.

This is why, when you gather together children without language in one place, they invent a language so quickly and so naturally—their brains already expect to do this.

Maybe we don't have a distinct "language organ" that is totally absent in other animals, but the human ability to absorb a language is nevertheless unmatched. It is very difficult to teach the simplest language to the smartest ape, but learning language is effortless for humans, like swimming for a dolphin.

Put another way, human language and the human brain are shaped not just by evolution, but by coevolution.[20] It is similar with flowering plants and their insect pollinators, both of which exploded in diversity within the past geological era. Plants developed aromatic smells, bright petals, and sweet nectar; insects acquired wings, proboscises, and color vision—and both benefited immensely as a result. It's not that either of them is adapted to the other—both are adapted to an interconnected life, and that is what makes them both so successful.

If you think of it this way, then you realize what makes humans special. It's not that there was some dividing line that separated us from other species. It's just that others might have flown at 11 kilometers per second, and we reached 12 kilometers per second. And at some point, our words stopped getting forgotten.

In that moment, we became something other than merely another species. We gave rise to a new form of life—like the word of god placed into the golem's mouth.

The Second Birth of Life

The word *life* can be understood in two different ways.

In the first sense, it describes the existence of an individual. This is the *life* we mean when we say, "my life." This life is always mortal.

In the second sense, *life* is the continued existence of a lineage.

This kind of *life* traverses generations, connecting them with a common thread.

It is thanks to this common thread that I have used the pronoun "we"—my favorite pronoun—to describe, throughout this book, living beings from the distant past.

This dichotomy—*we* and *I*—runs deep. If we look close enough, we can find it embedded in the very nature of molecules that make up living cells: DNA and proteins. In chapter 1, we saw that DNA can be copied but cannot do anything useful, whereas proteins can do everything except be copied. And so it is DNA that carries within itself an unbroken line of *we*, whereas proteins, which are discarded and re-created from scratch every time, only have their individual *I*'s.

In the broadest sense, *I*-life is the life of matter; *we*-life is the life of information.

Matter is the stuff from which everything is made; information is how the stuff is arranged.

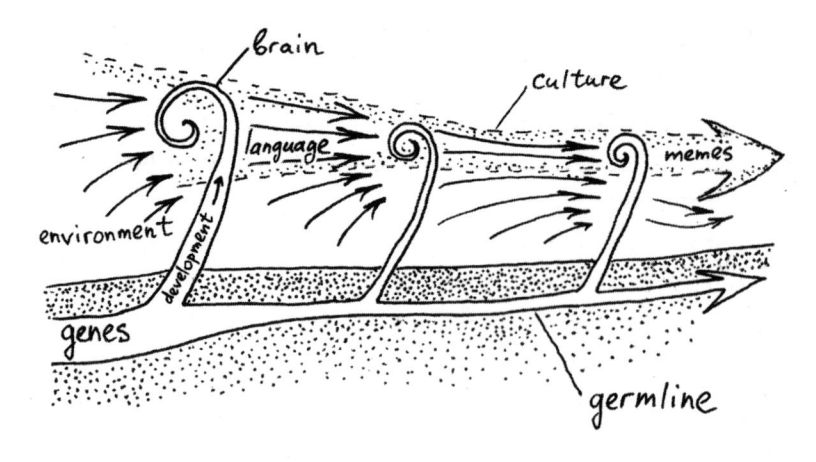

When I say that I am alive or that a dog is alive, I'm referring to the stuff, the material objects: the dog and I are considered alive because we can eat, breathe, move, and so on. That's *I*-life. But when I'm saying that the human species is alive or that cyanobacteria are alive or that life on Earth is alive, I'm referring to the arrangements: what is actually *alive* in this sense is not a human or a dog but *humanness* and *dogness*, not individual cyanobacteria, but the abilities and properties of those cyanobacteria, not specific DNA molecules made of specific atoms, but the order in which their letters are arranged in new organisms again and again and again.

Culture is also like that: it is its own separate *we*-life, sustained by repeatedly re-creating itself in new material objects—our brains.

At the most general level, culture is sustained imitation. Artists imitate famous painters. Musicians imitate conventions of a musical style. Almost all modern humans imitate each other in wearing clothes.

Culture is broader than language, but much of our cultural knowledge hinges on words. Doctors start their medical education by learning Latin anatomical terms. Lawyers imitate past justice by citing verbal accounts of precedent cases. Scientists imitate past ideas when they refer to other research in their papers. Language is itself an imitated skill and a conduit for the majority of other cultural knowledge that we imitate—from mathematics to funny pictures on the internet. (Think about it: does a dog understand why an internet meme is funny? It doesn't—you need language for that.)

This word *meme* is actually broader than funny pictures. It refers more generally to units of cultural information—*that which is imitated*. It's analogous to genes. Like genes, memes get passed on. Like genes, they evolve, adapting to their environment—the human mind.

So we can think of the birth of human language, which set forth an explosion of culture, as the birth of a new form of life sustained by imitation of memes rather than by replication of genes.

Billions of years ago, genes broke away from individual atoms and became subjects to eternity.

And so did culture transcend the confines of our individual, mortal bodies, a *we* that broke away from a sea of *I*'s.

This is the biggest difference between *we* and *I*: immortality. Every *I* has an end. But *we* does not have to.

That's why, I think, each of us yearns to be part of something bigger than us—whether it is family, friendship, fulfilling career, or religion. Mortality pushes us to participate in what we hope will exist forever, to share with others some eternal, unshakeable, everlasting *we*. You might not want to be immortal, but I bet you would like to have *legacy*.

And what will your legacy be?

Your genes, in a space of one generation, will be diluted in half. If your children have children, after two generations, only a quarter of your genetic *I* will be left for anyone to behold; in three, only an eighth; and in a couple hundred years, you will dissolve, as will we all, in the genetic river of the human species.

But your words?

Your words can live for as long as there's anyone there to hear them.

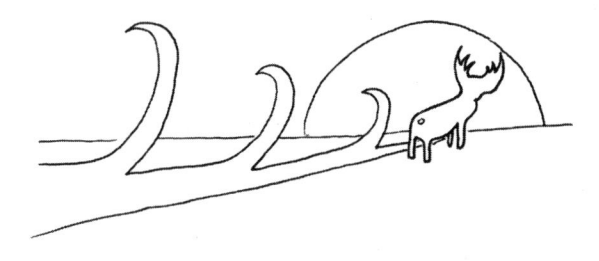

Epilogue

We began this book with a question: if everything in the world is made of the same stuff as science seems to suggest, then why does it feel so special to be me—on so many levels? It feels special to be alive in a world of nonlife. It feels special to be a human in a world of nonhumans. It feels special to look from my own eyes into the eyes of others. But any time we try to zero in on the specialness, to find the dividing line, we come up short.

Are these ideas, then, all imaginary? Throughout the book, I've argued that that's not the case: the ideas of nature, or essences, as we have called them, really do exist independently of our imagination.

But then as we zeroed in on *this* difference—imagination versus reality—we found that it also dissolves at close inspection. Our imagination *is* part of reality. A human idea is nothing more than a highly complex nature's idea.

So ironically it is in fact the human brain where nature's ideas—alive and not alive, human and nonhuman, first person and third person—express themselves most clearly. It just doesn't make them any less real.

In chapter 1, I used the example of aliens coming to our planet for a vodka tasting. I argued that although Russian wheat vodka and Polish potato vodka might be chemically the same, the aliens could still learn the difference if they analyzed the entire cultural context, rather than just the chemical composition of the two liquids—and so the difference does exist in reality rather than simply in our minds.

But where exactly should the aliens look for this "cultural context" that would distinguish the vodkas?

The quickest and most reliable way—even though it is far ahead of our human technology—would be to extract it from a brain of a human vodka connoisseur, such as myself. It is the place where the context is physically stored as a configuration of neurons and synapses.

This does not mean you could not understand the distinction in any other way. If the aliens spent years in Russia or Poland, they would eventually learn the distinction themselves by interpreting patterns of history and culture.

But my brain has already done the work of synthesizing everything necessary for making that distinction, including fluency in the Russian language, which the aliens would need to interpret the data. As long as aliens have a way to analyze the combined activity and configuration of every neuron and synapse in my brain, they can just abduct one person and not look any further.

And so it is with nature's ideas. They do exist outside of us—but inside our heads, they compress. Our minds are nature's idea labs, where patterns of the world accumulate, interact, fester—and from where they eventually break free in the form of culture.

This is what working with sea slugs made me understand. All ideas are essences. Even the words, *essence* and *idea*, really mean one thing—one is a translation, another a derivation of the Platonic *eidos*. In modern language, the meaning is *abstraction of a pattern*. It can happen outside of us, when a pattern of interactions between protons and electrons is abstracted in the chemical properties of carbon, setting the rules for life on Earth. It can happen inside of us, when we abstract the patterns of scientific knowledge accumulated over thousands of years and imagine this atom of carbon, embedding the essence within the geometry of our neural networks, making it an idea.

As time moves forward, the world moves from unified to diverse, from general to specific. The singularity of the Big Bang expands and gives rise to multiple physical forces, which give rise to a variety of chemical elements, which give rise to an infinite number of molecules, which—on planet Earth—give rise to ancestral organisms, which branch and branch and branch out into the diversity of life we know today. This upward branching tree is how the world grows, and our individual lives are its smallest branches.

But as these individual lives of ours move forward in time, our minds move in the reverse direction. As we learn about the world, we take in more and more specifics and form deeper and deeper abstractions to understand and predict them—an inverted tree, from specifics to generalities, and those into broader generalities.

As this inverted tree grows, combining branches into nodes, and then those nodes into further, bigger nodes, and so on to infinity, a singularity of abstraction, it reconstructs the history of the world in reverse.

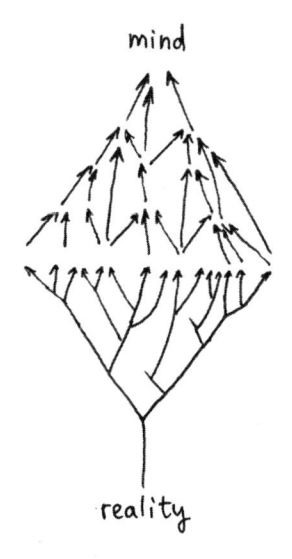

A child who knows nothing about evolution can tell that chickens and turkeys belong in one group, whereas mice and squirrels in a different one. The child sees consistent patterns within each pair of animals and links them into two separate mental nodes. But those mental nodes only form this way because millions of years ago, chickens and turkeys were one species, and squirrels and mice were another. And so, inadvertently, the child's brain reconstructs long-term patterns of evolution.

Some might even say that the brain is the *only* place where such categories exist—there's no such object as "bird" or "mammal," only specific birds and specific mammals.

But there's a better way to think of it. The human brain is where abstract ideas of nature—bird, mammal, life and nonlife, Russian vodka and Polish vodka—physically manifest in their most compressed form. And that is what makes each one of us special.

Here, then, is the answer to the question: you must stop seeing your mind as being opposed to nature and rather see it as an integral part of it. This makes you realize that *you are special precisely because you can think of it*—your brain is a corner of the universe where nature's ideas compress into a singularity, and it is only from the perspective of this singularity that you can ponder, understand, and appreciate your uniqueness.

The moment you let go of the line that separates imagination and reality is the moment you finally hear it: the sound of one hand clapping.

And you become nature.

Acknowledgments

I have always been tremendously lucky with my teachers, and I am thankful to everyone who has ever taught me anything. There are a few people I would like to mention specifically. Tatiana Viktorovna Selennova, who taught biology, and Tatiana Vasilyevna Martynova, who taught Russian literature, left the deepest impact on me out of all my high school teachers. At St. Petersburg State University, I am especially grateful to many people: our genetics professor Oleg Tikhodeev, who infected us with his passion for the link between genes and the brain and the mysteries of the evolutionary past; the zoology professor Andrei Granovich, who showed us a philosophical approach to minute anatomical features of various worms, which he drew on the blackboard with astounding speed using a dozen multicolored chalks; the botanists Maksim Baranov and Galina Borisovskaya, for their inspired imitations of plant organs ("I am a bud"); the physics professor Valentin Korotkov, whose thunderous declamations of Schrödinger's equation and charades about Maxwell's demon opened up the fabric of reality before our eyes; to Professor Natalia Chezhina, who taught us inorganic chemistry in the cavernous, dimly lit, Grand Chemistry Auditorium on Sredniy Prospect; the entomology professor Nikita Kluge, who taught us about insects in the woods by the Svir River; and similarly to Dmitry Aristov, who oversaw our practice at the White Sea, where we spent days gathering bucketfuls of bizarre sea creatures and nights staring into microscopes, trying and failing to draw the creatures under the midnight sun of the near-Arctic summer. From the biochemist Georgiy Volskiy I first heard about *Aplysia*, my future research object, and his lectures on bioenergetics were the foundation for my entire understanding of metabolism. The immunology professor Aleksandr Polevschikov lectured like a Roman orator before a crowd, with passion and fervor: in his telling, inflammation caused by a splinter turned into an action thriller. In Aleksei Baskakov's lectures on cell biology, molecules and even phosphorylation sites came alive, like strange cartoon characters. In fact, many cartoon characters in this book, like the deer with a black eye, made their debut as doodles in my notes from that class. (I know the deer looks more like a moose, but one does not question evolution.)

I am no less grateful to my English-speaking teachers. My adviser at Oxford, the great, proud Terry Butters, taught me the order and discipline of scientific work, despite the unmentionable quantities of English ale that our lab consumed on a weekly basis. The most important thing that Terry did for me, however, was to introduce me to Katy, my future wife. To Fred Goldberg, my mentor at Harvard, I am grateful for his intellectual intensity—I had never had to use as much brain power as I did in his lab. To Tom Carew, my current boss, I owe a fascination with the most basic building blocks of the brain. I am also grateful to Tom for always letting me be myself: asking strange questions and moving in whichever direction they take me.

At one point when working in Boston, I had to return to Russia to renew my US visa and got stuck for two months. On the first day of this unintended vacation, I reached out to Asya Kazantseva, my old friend and former college classmate, a popular science celebrity in Russia. We had been working on her first book, *Who Would Have Thought*. I drew some doodles for her (ironically, my career as an illustrator preceded writing, although, as you can tell, I have no idea how to draw anything.) I had also been sporadically blogging about biology, and Asya suggested that I try publishing some popular science while I had free time. She connected me to Tania Cohen, at the time editor in chief of the great Metropol (later, Knife Media). As a result, in the next few years I wrote hundreds of popular science articles in various Russian media, first in Metropol, then in lots of other places. I am grateful to everyone who read or published me, but most of all to Asya and Tania, without whom I would have never thought of creating anything like this book. They taught me to write in a way that people can understand—and that, in turn, taught me to think that way too.

Special thanks go to my Russian publisher, Alpina Non-Fiction (ANF). I would like to especially thank ANF's editor in chief, Pavel Podkosov, as well as Marina Krasavina, Aleksandra Shuvalova, Valentina Bologova, Maria Sirotina, Natalia Pepelina, and everyone else who worked on the book. I am also especially grateful to my science editor, Sergei Yastrebov, for his unrelenting thoughtfulness and for saving me from several obvious mistakes.

My book had a long journey on its way to the English version. Originally ANF was planning to publish it through its international division. But the project died in February 2022, when Russia invaded Ukraine. After several months during which book rights were the last thing I was thinking about, I eventually negotiated the transfer of translation rights to me (since mail still does not travel between the United States and Russia, the operation involved passing documents through multiple intermediaries in Austria and Portugal, including—I am not kidding—a meeting in a café in Vienna).

I would have no idea what to do with those rights had I not stumbled upon my agent, Jaime Marshall. I still cannot believe this singular stroke of luck I was handed—it really ranks on the scale of getting accepted to Oxford and—I don't know—being born. Jaime guided me through the entire process of publication, from book proposal to reworking the book for the English audience. I am also thankful to my publicists, Jess Pellien and Kailey Tse-Harlow, for helping me get out of my dark office and start talking to the world.

I am thankful to my publishers, Prometheus (in the United States) and Swift Press (in the United Kingdom). In the United States, I want to especially thank Jake Bonar, Alden Perkins, Erin McGarvey, Jason Rossi, Emily Jeffers, Piper Wallis, Neil Cotterill, and Justine Connelly; in the United Kingdom, Mark Richards and Ruth Killick, and everyone else who was involved in making this book a reality.

I am grateful to all my friends from whom I borrowed many ideas, and in particular Joha Coludar and Misha Kostylev, each of whom influenced me unmeasurably.

Many topics in this book were first developed, tested, and polished in my life science lectures at New York University's Liberal Studies. For years, the book and the class had a symbiotic relationship—ideas migrated in both directions, and as a result both the class and the book became richer in material and stronger in argumentation. I would like to thank my students, the test subjects in these not-always-successful attempts to explain all of biology in one semester. I also want to thank my colleagues at Liberal Studies, whose multidisciplinary mindset and global perspective were a major inspiration for the book.

As an honest mammal, I owe everything that is good in my life to my family. Without it, I would be a different person. This book also would have been completely different and most likely would not have existed at all. I want to thank my parents for a happy childhood and, frankly, adulthood as well. I am grateful to my mom for passing on her hunger to understand anything I lay my eyes on; to my dad, for his love of language and clarity of words; to my babushka, for her hooligan tendencies; and to my wife, for her sense of responsibility for the future. Also for forgiving me the endless ruined weekends, glassy eyes, and thousands of words in a language she did not know. Instead of guilt trips, I somehow always got food and occasionally a can of beer. That is worth living for.

Notes

Chapter 1: In the Beginning Were the Letters

1 Cameron, A. G. W., "Abundances of the Elements in the Solar System," *Space Science Reviews* 15 (1973): 121–46.

2 UniProt Consortium, "UniProt: The Universal Protein Resource," *Nucleic Acids Research* 45 (2017): D158–69.

3 Orgel, L. E., "Prebiotic Chemistry and the Origin of the RNA World," *Critical Reviews in Biochemistry and Molecular Biology* 39 (2004): 99–123; Wolf, Y. I., and Koonin, E. V., "On the Origin of the Translation System and the Genetic Code in the RNA World by Means of Natural Selection, Exaptation, and Subfunctionalization," *Biology Direct* 2 (2007): 14.

4 Lincoln, T. A., and Joyce, G. F., "Self-Sustained Replication of an RNA Enzyme," *Science* 323 (2009): 1229–32; and Robertson, M. P., and Joyce, G. F., "Highly Efficient Self-Replicating RNA Enzymes," *Chemistry & Biology* 21 (2014): 238–45.

5 Cafferty, B. J., Fialho, D. M., Khanam, J., Krishnamurthy, R., and Hud, N. V., "Spontaneous Formation and Base Pairing of Plausible Prebiotic Nucleotides in Water," *Nature Communications* 7 (2016): 11328.

6 Martin, W., Baross, J., Kelley, D., and Russell, M. J., "Hydrothermal Vents and the Origin of Life," *Nature Reviews Microbiology* 6 (2008): 805–14; Koonin, E. V., *The Logic of Chance: The Nature and Origin of Biological Evolution* (Pearson Education, 2011); Sojo, V., Herschy, B., Whicher, A., Camprubi, E., and Lane, N., "The Origin of Life in Alkaline Hydrothermal Vents," *Astrobiology* 16, no. 2 (2016): 181–97.

7 Baaske, P., Weinert, F. M., Duhr, S., Lemke, K. H., Russell, M. J., and Braun, D., "Extreme Accumulation of Nucleotides in Simulated Hydrothermal Pore Systems," *Proceedings of the National Academy of Sciences*, 104, no. 23 (2007): 9346–51; Burcar, B. T., Barge, L. M., Trail, D., Watson, E. B., Russell, M. J., and McGown, L. B., "RNA Oligomerization in Laboratory Analogues of Alkaline Hydrothermal Vent Systems," *Astrobiology* 15, no. 7 (2015): 509–22; Mulkidjanian, A. Y., "On the Origin of Life in the Zinc World: 1. Photosynthesizing, Porous Edifices Built of Hydrothermally Precipitated Zinc Sulfide as Cradles of Life on Earth," *Biology Direct* 4 (2009): 26; Mulkidjanian, A. Y., and Galperin, M. Y., "On the Origin of Life in the Zinc World. 2. Validation of the Hypothesis on the Photosynthesizing Zinc Sulfide Edifices as Cradles of Life on Earth," *Biology Direct* 4 (2009): 27.

8 Kelley, D. S., Karson, J. A., Fruh-Green, G. L., et al., "A Serpentinite-Hosted Ecosystem: The Lost City Hydrothermal Field," *Science* 307, no. 5714 (2005): 1428–34;

Martin, W., Baross, J., Kelley, D., and Russell, M. J., "Hydrothermal Vents and the Origin of Life," *Nature Reviews Microbiology* 6 (2008): 805–14; Proskurowski, G., Lilley, M. D., Seewald, J. S., et al., "Abiogenic Hydrocarbon Production at Lost City Hydrothermal Field," *Science* 319, no. 5863 (2008): 604–7.

Chapter 2: A Good Idea

1 Lucas, J. R., "Wilberforce and Huxley: A Legendary Encounter," *Historical Journal* 22, no. 2 (1979): 313–30.

2 Barlow, D., "The Devil Within: Evolution of a Tragedy," *Weather* 52, no. 11 (1997): 337–41.

3 MacCallum, R. M., Mauch, M., Burt, A., and Leroi, A. M., "Evolution of Music by Public Choice," *Proceedings of the National Academy of Sciences* 109, no. 30 (2012): 12081–86.

4 Dobzhansky, T., "Nothing in Biology Makes Sense except in the Light of Evolution," *American Biology Teacher* 75, no. 2 (2013): 87–91.

5 Dawkins, R. *The Selfish Gene* (Oxford University Press, 1976).

Chapter 3: The Birth of Complexity

1 Woese, C. R., Kandler, O., and Wheelis, M. L., "Towards a Natural System of Organisms: Proposal for the Domains Archaea, Bacteria, and Eucarya," *Proceedings of the National Academy of Sciences*, 87, no. 12 (1990): 4576–79.

2 Oliver, T., Sánchez-Baracaldo, P., Larkum, A. W., Rutherford, A. W., and Cardona, T., "Time-Resolved Comparative Molecular Evolution of Oxygenic Photosynthesis," *Biochimica et Biophysica Acta (BBA)-Bioenergetics* 1862, no. 6 (2021): 148400.

3 Xiong, J., Fischer, W. M., Inoue, K., Nakahara, M., and Bauer, C. E., "Molecular Evidence for the Early Evolution of Photosynthesis," *Science* 289, no. 5485 (2000): 1724–30; Des Marais, D. J., "When Did Photosynthesis Emerge on Earth?" *Science* 289, no. 5485 (2000): 1703–5.

4 Crowe, S. A., Jones, C., Katsev, S., et al., "Photoferrotrophs Thrive in an Archean Ocean Analogue," *Proceedings of the National Academy of Sciences* 105, no. 41 (2008): 15938–43.

5 Olson, J. M., "Photosynthesis in the Archean Era," *Photosynthesis Research* 88 (2006): 109–17.

6 Johnston, D. T., Wolfe-Simon, F., Pearson, A., and Knoll, A. H., "Anoxygenic Photosynthesis Modulated Proterozoic Oxygen and Sustained Earth's Middle Age," *Proceedings of the National Academy of Sciences* 106, no. 40 (2009): 16925–29.

7   Margulis, L., and Sagan, D., *Microcosmos: Four Billion Years of Microbial Evolution* (University of California Press, 1997).

8   Pedersen, R. B., Rapp, H. T., Thorseth, I. H., et al., "Discovery of a Black Smoker Vent Field and Vent Fauna at the Arctic Mid-Ocean Ridge," *Nature Communications* 1, no. 1 (2010): 126.

9   Seitz, K. W., Lazar, C. S., Hinrichs, K. U., Teske, A. P., and Baker, B. J., "Genomic Reconstruction of a Novel, Deeply Branched Sediment Archaeal Phylum with Pathways for Acetogenesis and Sulfur Reduction," *ISME Journal* 10, no. 7 (2016): 1696–1705.

10 Zaremba-Niedzwiedzka, K., Caceres, E. F., Saw, J. H., et al., "Asgard Archaea Illuminate the Origin of Eukaryotic Cellular Complexity," *Nature* 541, no. 7637 (2017): 353–58.

11 Eme, L., Tamarit, D., Caceres, E. F., et al., "Inference and Reconstruction of the Heimdallarchaeial Ancestry of Eukaryotes," *Nature* 618, no. 7967 (2023): 992–99.

12 Imachi, H., Nobu, M. K., Nakahara, N., et al., "Isolation of an Archaeon at the Prokaryote–Eukaryote Interface," *Nature* 577, no. 7791 (2020): 519–25.

13 Shiratori, T., Suzuki, S., Kakizawa, Y., and Ishida, K. I., et al., "Phagocytosis-Like Cell Engulfment by a Planctomycete Bacterium," *Nature Communications* 10, no. 1 (2019): 5529.

14 Forsman, K., CartoonStock, June 25, 2014, https://www.cartoonstock.com/cartoonview.asp?catref=kfon351.

15 Oeppen, J., and Vaupel, J. W., "Broken Limits to Life Expectancy," *Science* 296, no. 5570 (2002): 1029–31.

16 World Health Organization, *Antimicrobial Resistance: Global Report on Surveillance* (World Health Organization, 2014).

17 Harris, J. J., Jolivet, R., and Attwell, D., "Synaptic Energy Use and Supply," *Neuron* 75, no. 5 (2012): 762–77; Howarth, C., Gleeson, P., and Attwell, D., "Updated Energy Budgets for Neural Computation in the Neocortex and Cerebellum," *Journal of Cerebral Blood Flow & Metabolism* 32, no. 7 (2012): 1222–32.

Chapter 4: When All Else Fails

1   González, A., Hall, M. N., Lin, S. C., and Hardie, D. G., "AMPK and TOR: The Yin and Yang of Cellular Nutrient Sensing and Growth Control," *Cell Metabolism* 31, no. 3 (2020): 472–92.

2   Kukushkin, N. V., Tabassum, T., and Carew, T. J., "Precise Timing of ERK Phosphorylation/Dephosphorylation Determines the Outcome of Trial Repetition during

Long-Term Memory Formation," *Proceedings of the National Academy of Sciences* 119, no. 40 (2022): e2210478119.

3 Brewster, J. L., de Valoir, T., Dwyer, N. D., Winter, E., and Gustin, M. C., "An Osmosensing Signal Transduction Pathway in Yeast," *Science* 259, no. 5102 (1993): 1760–63.

4 Lehtonen, J., Kokko, H., and Parker, G. A., "What Do Isogamous Organisms Teach Us about Sex and the Two Sexes?" *Philosophical Transactions of the Royal Society B: Biological Sciences* 371, no. 1706 (2016): 20150532.

Chapter 5: The Moving Kingdom

1 Berner, R. A., "The Rise of Plants and Their Effect on Weathering and Atmospheric CO2," *Science* 276, no. 5312 (1997): 544–46.

2 Bar-On, Y. M., Phillips, R., and Milo, R., "The Biomass Distribution on Earth," *Proceedings of the National Academy of Sciences* 115, no. 25 (2018): 6506–11.

3 Forterre, Y., Skotheim, J. M., Dumais, J., and Mahadevan, L., "How the Venus Flytrap Snaps," *Nature* 433, no. 7024 (2005): 421–25.

4 Patek, S. N., Korff, W. L., and Caldwell, R. L., "Deadly Strike Mechanism of a Mantis Shrimp," *Nature* 428, no. 6985 (2004): 819–20.

5 Ryan, J. F., Pang, K., Schnitzler, C. E., et al., "The Genome of the Ctenophore Mnemiopsis Leidyi and Its Implications for Cell Type Evolution," *Science* 342, no. 6164 (2013): 1242592; Moroz, L. L., Kocot, K. M., Citarella, M. R., et al., "The Ctenophore Genome and the Evolutionary Origins of Neural Systems," *Nature* 510, no. 7503 (2014): 109–14; Simion, P., Philippe, H., Baurain, D., et al., "A Large and Consistent Phylogenomic Dataset Supports Sponges as the Sister Group to All Other Animals," *Current Biology* 27, no. 7 (2017): 958–67.

6 Philippe, H., Derelle, R., Lopez, P., et al., "Phylogenomics Revives Traditional Views on Deep Animal Relationships," *Current Biology* 19, no. 8 (2009): 706–12; Nielsen, C., "Six Major Steps in Animal Evolution: Are We Derived Sponge Larvae?" *Evolution & Development* 10, no. 2 (2008): 241–57; Cavalier-Smith, T., "Origin of Animal Multicellularity: Precursors, Causes, Consequences—The Choanoflagellate/Sponge Transition, Neurogenesis and the Cambrian Explosion," *Philosophical Transactions of the Royal Society B: Biological Sciences* 372, no. 1713 (2017): 20150476.

7 Cavalier-Smith, T., "Origin of Animal Multicellularity: Precursors, Causes, Consequences—The Choanoflagellate/Sponge Transition, Neurogenesis and the Cambrian Explosion," *Philosophical Transactions of the Royal Society B: Biological Sciences* 372, no. 1713 (2017): 20150476.

8 Fairclough, S. R., Dayel, M. J., and King, N., "Multicellular Development in a Choanoflagellate," *Current Biology* 20, no. 20 (2010): R875–76.

9   Leys, S. P., and Degnan, B. M., "Cytological Basis of Photoresponsive Behavior in a Sponge Larva," *The Biological Bulletin* 201, no. 3 (2001): 323–38.

10 Nielsen, C., "Six Major Steps in Animal Evolution: Are We Derived Sponge Larvae?" *Evolution & Development* 10, no. 2 (2008): 241–57; Wörheide, G., Dohrmann, M., Erpenbeck, D., et al., "Deep Phylogeny and Evolution of Sponges (Phylum Porifera)," *Advances in Marine Biology* 61 (2012): 1–78.

11 Lee, W. L., Reiswig, H. M., Austin, W. C., and Lundsten, L., "An Extraordinary New Carnivorous Sponge, Chondrocladia Lyra, in the New Subgenus Symmetrocladia (Demospongiae, Cladorhizidae), from off of Northern California, USA," *Invertebrate Biology* 131, no. 4 (2012): 259–84.

12 Collins, A. G., "Phylogeny of Medusozoa and the Evolution of Cnidarian Life Cycles," *Journal of Evolutionary Biology* 15, no. 3 (2002): 418–32.

13 Steinmetz, P. R., Kraus, J. E., Larroux, C., et al., "Independent Evolution of Striated Muscles in Cnidarians and Bilaterians," *Nature* 487, no. 7406 (2012): 231–34.

14 Satterlie, R. A., "Do Jellyfish Have Central Nervous Systems?" *Journal of Experimental Biology* 214, no. 8 (2011): 1215–23.

15 Nielsen, C., "Six Major Steps in Animal Evolution: Are We Derived Sponge Larvae?" *Evolution & Development* 10, no. 2 (2008): 241–57.

16 Kallmeyer, J., Pockalny, R., Adhikari, R. R., Smith, D. C., and D'Hondt, S., "Global Distribution of Microbial Abundance and Biomass in Subseafloor Sediment," *Proceedings of the National Academy of Sciences* 109, no. 40 (2012): 16213–16.

17 Narbonne, G. M., "The Ediacara Biota: Neoproterozoic Origin of Animals and Their Ecosystems," *Annual Review of Earth and Planetary Science* 33, no. 1 (2005): 421–42.

18 Budd, G. E., and Jensen, S., "A Critical Reappraisal of the Fossil Record of the Bilaterian Phyla," *Biological Reviews* 75, no. 2 (2000): 253–95.

19 Zhang, X., and Cui, L., "Oxygen Requirements for the Cambrian Explosion," *Journal of Earth Science* 27 (2016): 187–95; Mills, D. B., and Canfield, D. E., "Oxygen and Animal Evolution: Did a Rise of Atmospheric Oxygen 'Trigger' the Origin of Animals?" *BioEssays* 36, no. 12 (2014): 1145–55; Sperling, E. A., Frieder, C. A., Raman, A. V., Girguis, P. R., Levin, L. A., and Knoll, A. H., "Oxygen, Ecology, and the Cambrian Radiation of Animals," *Proceedings of the National Academy of Sciences* 110, no. 33 (2013): 13446–51.

20 Cavalier-Smith, T., "Cell Evolution and Earth History: Stasis and Revolution," *Philosophical Transactions of the Royal Society B: Biological Sciences* 361, no. 1470 (2006): 969–1006; Cavalier-Smith, T., "Origin of Animal Multicellularity: Precursors, Causes, Consequences—The Choanoflagellate/Sponge Transition, Neurogenesis and the Cambrian Explosion," *Philosophical Transactions of the Royal Society B: Biological Sciences* 372, no. 1713 (2017): 20150476.

21 Mángano, M. G., and Buatois, L. A., "Decoupling of Body-Plan Diversification and Ecological Structuring during the Ediacaran–Cambrian Transition: Evolutionary and Geobiological Feedbacks," *Proceedings of the Royal Society B: Biological Sciences* 281, no. 1780 (2014): 20140038.

22 Holland, P. W., "Did Homeobox Gene Duplications Contribute to the Cambrian Explosion?" *Zoological Letters* 1 (2015): 1–8.

23 Ereskovsky, A. V., Borchiellini, C., Gazave, E., et al., "The Homoscleromorph Sponge Oscarella Lobularis, a Promising Sponge Model in Evolutionary and Developmental Biology: Model Sponge Oscarella Lobularis," *BioEssays* 31, no. 1 (2009): 89–97.

Chapter 6: Land!

1 Caldwell, J. P., Thorp, J. H., and Jervey, T. O., "Predator-Prey Relationships among Larval Dragonflies, Salamanders, and Frogs," *Oecologia* 46 (1980): 285–89.

2 Watanabe, Y., Martini, J. E., and Ohmoto, H., "Geochemical Evidence for Terrestrial Ecosystems 2.6 Billion Years Ago," *Nature* 408, no. 6812 (2000): 574–78.

3 Honegger, R., Edwards, D., and Axe, L., "The Earliest Records of Internally Stratified Cyanobacterial and Algal Lichens from the Lower Devonian of the Welsh Borderland," *New Phytologist* 197, no. 1 (2013): 264–75; Heckman, D. S., Geiser, D. M., Eidell, B. R., Stauffer, R. L., Kardos, N. L., and Hedges, S. B., "Molecular Evidence for the Early Colonization of Land by Fungi and Plants," *Science* 293, no. 5532 (2001): 1129–33.

4 Sancho, L. G., De la Torre, R., Horneck, G., et al., "Lichens Survive in Space: Results from the 2005 LICHENS Experiment," *Astrobiology* 7, no. 3 (2007): 443–54.

5 Delwiche, C. F., & Cooper, E. D., "The Evolutionary Origin of a Terrestrial Flora," *Current Biology* 25, no. 19 (2015): R899–R910.

6 Field, K. J., Pressel, S., Duckett, J. G., Rimington, W. R., and Bidartondo, M. I., "Symbiotic Options for the Conquest of Land," *Trends in Ecology & Evolution* 30, no. 8 (2015): 477–86.

7 Kroken, S. B., Graham, L. E., and Cook, M. E., "Occurrence and Evolutionary Significance of Resistant Cell Walls in Charophytes and Bryophytes," *American Journal of Botany* 83, no. 10 (1996): 1241–54.

8 Wellman, C. H., "Origin, Function and Development of the Spore Wall in Early Land Plants," in *The Evolution of Plant Physiology* (Academic Press, 2004), 43–63.

9 Harrison, C. J., and Morris, J. L., "The Origin and Early Evolution of Vascular Plant Shoots and Leaves," *Philosophical Transactions of the Royal Society B: Biological Sciences* 373, no. 1739 (2018): 20160496.

10 Beerling, D. J., "Atmospheric Carbon Dioxide: A Driver of Photosynthetic Eukaryote Evolution for over a Billion Years?" *Philosophical Transactions of the Royal Society B: Biological Sciences* 367, no. 1588 (2012): 477–82; Dorrell, R. G., and Smith, A. G., "Do Red and Green Make Brown? Perspectives on Plastid Acquisitions within Chromalveolates," *Eukaryotic Cell*, 10, no. 7 (2011): 856–68.

11 Rota-Stabelli, O., Daley, A. C., and Pisani, D., "Molecular Timetrees Reveal a Cambrian Colonization of Land and a New Scenario for Ecdysozoan Evolution," *Current Biology* 23, no. 5 (2013): 392–98.

12 Bar-On, Y. M., Phillips, R., and Milo, R., "The Biomass Distribution on Earth," *Proceedings of the National Academy of Sciences* 115, no. 25 (2018): 6506–11.

13 Clack, J. A., "Gaining Ground," in *The Origin and Evolution of Tetrapods* (Indiana University Press, 2002).

14 Verberk, W. C., and Bilton, D. T., "Can Oxygen Set Thermal Limits in an Insect and Drive Gigantism?" *PLoS One* 6, no. 7 (2011): e22610; Clapham, M. E., and Karr, J. A., "Environmental and Biotic Controls on the Evolutionary History of Insect Body Size," *Proceedings of the National Academy of Sciences* 109, no. 27 (2012): 10927–30.

15 Pittman, R. N., *Regulation of Tissue Oxygenation* (Biota Publishing, 2016).

16 Rota-Stabelli, O., Daley, A. C., and Pisani, D., "Molecular Timetrees Reveal a Cambrian Colonization of Land and a New Scenario for Ecdysozoan Evolution," *Current Biology* 23, no. 5 (2013): 392–98.

Chapter 7: When the World Ends

1 Clack, J. A., "Gaining Ground: The Evolution of Terrestriality," in *Gaining Ground: The Origin and Evolution of Tetrapods* (Indiana University Press, 2002).

2 Peyser, C. E., and Poulsen, C. J., "Controls on Permo-Carboniferous Precipitation over Tropical Pangaea: A GCM Sensitivity Study," *Palaeogeography, Palaeoclimatology, Palaeoecology* 268, nos. 3–4 (2008): 181–92; Dunne, E. M., Close, R. A., Button, D. J., et al., "Diversity Change during the Rise of Tetrapods and the Impact of the 'Carboniferous Rainforest Collapse,'" *Proceedings of the Royal Society B: Biological Sciences* 285, no. 1872 (2018): 20172730.

3 DeMar, R., and Barghusen, H. R., "Mechanics and the Evolution of the Synapsid Jaw," *Evolution* 26, no. 4 (1972): 622–37.

4 Sidor, C. A., and Hopson, J. A. "Ghost Lineages and 'Mammalness': Assessing the Temporal Pattern of Character Acquisition in the Synapsida," *Paleobiology* 24, no. 2 (1998): 254–73.

5 Sumida, S., and Martin, K. L., eds., *Amniote Origins: Completing the Transition to Land* (Elsevier, 1997).

6   Smant, G., Stokkermans, J. P., Yan, Y., et al., "Endogenous Cellulases in Animals: Isolation of β-1, 4-Endoglucanase Genes from Two Species of Plant-Parasitic Cyst Nematodes," *Proceedings of the National Academy of Sciences* 95, no. 9 (1998): 4906–11; Watanabe, H., and Tokuda, G., "Animal Cellulases," *Cellular and Molecular Life Sciences CMLS* 58 (2001): 1167–78; Lo, N., Watanabe, H., and Sugimura, M., "Evidence for the Presence of a Cellulase Gene in the Last Common Ancestor of Bilaterian Animals," *Proceedings of the Royal Society of London. Series B: Biological Sciences* 270, no. S1 (2003): S69–72.

7   Shen, S. Z., Crowley, J. L., Wang, Y., et al., "Calibrating the End-Permian Mass Extinction," *Science* 334, no. 6061 (2011): 1367–72.

8   Clarkson, M. O., Kasemann, S. A., Wood, R. A., et al., "Ocean Acidification and the Permo-Triassic Mass Extinction," *Science* 348, no. 6231 (2015): 229–32.

9   Erwin, D. H., "The Permo-Triassic Extinction," *Nature* 367, no. 6460 (1994): 231–36.

10 Rothman, D. H., Fournier, G. P., French, K. L., et al., "Methanogenic Burst in the End-Permian Carbon Cycle," *Proceedings of the National Academy of Sciences* 111, no. 15 (2014): 5462–67.

11 Newitz, A., "Lystrosaurus: The Most Humble Badass of the Triassic," *National Geographic* May 28, 2013, https://www.nationalgeographic.com/science/phenomena/2013/05/28/lystrosaurus-the-most-humble-badass-of-the-triassic/; Botha, J., and Smith, R. M., "Lystrosaurus Species Composition across the Permo-Triassic Boundary in the Karoo Basin of South Africa," *Lethaia* 40, no. 2 (2007): 125–37.

12 Botha-Brink, J., "Burrowing in Lystrosaurus: Preadaptation to a Postextinction Environment?" *Journal of Vertebrate Paleontology* 37, no. 5 (2017): e1365080.

13 Conway, J., Kosemen, C. M., Naish, D., and Hartman, S., *All yesterdays: Unique and Speculative Views of Dinosaurs and Other Prehistoric Animals* (Irregular Books, 2013).

14 Choo, B., "*Jurassic Art: How Our Vision of Dinosaurs Has Evolved over Time*," The Conversation, June 16, 2015, https://theconversation.com/jurassic-art-how-our-vision-of-dinosaurs-has-evolved-over-time-42998/.

15 Alexander, R. M., "Dinosaur Biomechanics," *Proceedings of the Royal Society B: Biological Sciences* 273, no. 1596 (2006): 1849–55.

16 Vinther, J., "The True Colors of Dinosaurs," *Scientific American* 316, no. 3 (2017): 50–57.

17 Brett-Surman, M. K., Holtz, T. R., and Farlow, J. O., eds., *The Complete Dinosaur* (Indiana University Press, 2012).

18 Amiot, R., Lécuyer, C., Buffetaut, E., Escarguel, G., Fluteau, F., and Martineau, F., "Oxygen Isotopes from Biogenic Apatites Suggest Widespread Endothermy in Cretaceous Dinosaurs," *Earth and Planetary Science Letters* 246, no. 1–2 (2006): 41–54.

19 Varricchio, D. J., Moore, J. R., Erickson, G. M., Norell, M. A., Jackson, F. D., and Borkowski, J. J., "Avian Paternal Care Had Dinosaur Origin," *Science* 322, no. 5909 (2008): 1826–28; Meng, Q., Liu, J., Varricchio, D. J., Huang, T., and Gao, C., "Parental Care in an Ornithischian Dinosaur," *Nature* 431, no. 7005 (2004): 145–46.

20 Hearn, L., and Williams, A. C. D. C., "Pain in Dinosaurs: What Is the Evidence?" *Philosophical Transactions of the Royal Society B*, 374, no. 1785 (2019): 20190370.

21 Horner, J. R., "The Nesting Behavior of Dinosaurs," *Scientific American*, 250, no. 4 (1984): 130–37.

22 Xu, X., Zhou, Z., and Wang, X., "The Smallest Known Non-Avian Theropod Dinosaur," *Nature* 408, no. 6813 (2000): 705–8.

23 Riede, T., Eliason, C. M., Miller, E. H., Goller, F., and Clarke, J. A., "Coos, Booms, and Hoots: The Evolution of Closed-Mouth Vocal Behavior in Birds," *Evolution* 70, no. 8 (2016): 1734–46.

24 Pan, Y., Zheng, W., Sawyer, R. H., et al., "The Molecular Evolution of Feathers with Direct Evidence from Fossils," *Proceedings of the National Academy of Sciences* 116, no. 8 (2019): 3018–23.

25 Lautenschlager, S., Witmer, L. M., Altangerel, P., and Rayfield, E. J., "Edentulism, Beaks, and Biomechanical Innovations in the Evolution of Theropod Dinosaurs," *Proceedings of the National Academy of Sciences* 110, no. 51 (2013): 20657–62.

26 Kemp, T. S., "The Origin and Early Radiation of the Therapsid Mammal-Like Reptiles: A Palaeobiological Hypothesis," *Journal of Evolutionary Biology* 19, no. 4 (2006): 1231–47; Huttenlocker, A. K., "Body Size Reductions in Nonmammalian Eutheriodont Therapsids (Synapsida) during the End-Permian Mass Extinction," *PLoS One* 9, no. 2 (2014): e87553.

27 Berner, R. A., "The Carbon and Sulfur Cycles and Atmospheric Oxygen from Middle Permian to Middle Triassic," *Geochimica et Cosmochimica Acta* 69, no. 13 (2005): 3211–17.

28 Schmidt-Nielsen, K., *Animal Physiology: Adaptation and Environment* (Cambridge University Press, 1997).

29 Schmidt-Nielsen, K., *Animal Physiology: Adaptation and Environment* (Cambridge University Press, 1997).

30 Tucker, V. A., "Respiratory Physiology of House Sparrows in Relation to High-Altitude Flight," *Journal of Experimental Biology* 48, no. 1 (1968), 55–66.

31 Farmer, C. G., "The Evolution of Unidirectional Pulmonary Airflow," *Physiology* 30, no. 4 (2015): 260–72.

32 Farmer, C. G., and Sanders, K., "Unidirectional Airflow in the Lungs of Alligators," *Science* 327, no. 5963 (2010): 338–40.

33 Schachner, E. R., Cieri, R. L., Butler, J. P., and Farmer, C. G., "Unidirectional Pulmonary Airflow Patterns in the Savannah Monitor Lizard," *Nature* 506, no. 7488 (2014): 367–70.

34 Carrier, D. R. "The Evolution of Locomotor Stamina in Tetrapods: Circumventing a Mechanical Constraint," *Paleobiology* 13, no. 3 (1987): 326–41.

35 Charles-Dominique, P., "Nocturnality and Diurnality: An Ecological Interpretation of These Two Modes of Life by an Analysis of the Higher Vertebrate Eye," in *Phylogeny of the Primates: A Multidisciplinary Approach*, ed. W. P. Luckett and F. S. Szalay (Springer, 1975), 69–88.

36 Heesy, C. P., and Hall, M. I., "The Nocturnal Bottleneck and the Evolution of Mammalian Vision," *Brain Behavior and Evolution* 75, no. 3 (2010): 195–203.

37 Seebacher, F., "Dinosaur Body Temperatures: The Occurrence of Endothermy and Ectothermy," *Paleobiology* 29, no. 1 (2003): 105–22.

38 McNab, B. K., "The Evolution of Endothermy in the Phylogeny of Mammals," *The American Naturalist* 112, no. 983 (1978): 1–21.

39 Hillenius, W. J., "Turbinates in Therapsids: Evidence for Late Permian Origins of Mammalian Endothermy," *Evolution* 48, no. 2 (1994): 207–29.

40 Luo, Z. X., "Transformation and Diversification in Early Mammal Evolution," *Nature* 450 (2007): 1011–19.

41 Bennett, A. F., and Ruben, J. A., "Endothermy and Activity in Vertebrates," *Science* 206, no. 4419 (1979): 649–54.

42 Edelman, I. S., "Transition from the Polikilotherm to the Homeotherm: Possible Role of Sodium Transport and Thyroid Hormone," *Federation Proceedings* 35, no. 10 (August 1976): 2180–84; Else, P. L., Windmill, D. J., and Markus, V., "Molecular Activity of Sodium Pumps in Endotherms and Ectotherms," *American Journal of Physiology—Regulatory, Integrative and Comparative Physiology* 271, no. 5 (1996): R1287–94.

43 Hughes, D. A., Jastroch, M., Stoneking, M., and Klingenspor, M., "Molecular Evolution of UCP1 and the Evolutionary History of Mammalian Non-Shivering Thermogenesis," *BMC Evolutionary Biology* 9 (2009): 1–13.

44 Else, P. L., Windmill, D. J., and Markus, V., "Molecular Activity of Sodium Pumps in Endotherms and Ectotherms," *American Journal of Physiology—Regulatory, Integrative and Comparative Physiology* 271, no. 5 (1996): R1287–94.

45 Edelman, I. S., "Transition from the Polikilotherm to the Homeotherm: Possible Role of Sodium Transport and Thyroid Hormone," *Federation Proceedings* 35, no. 10 (August 1976): 2180–84.

46 Schmidt-Nielsen, K., *Animal Physiology: Adaptation and Environment* (Cambridge University Press, 1997).

47 Morgan, J. V., Gulick, S. P., Bralower, T., et al., "The Formation of Peak Rings in Large Impact Craters," *Science* 354, no. 6314 (2016): 878–82; Kring, D. A., Claeys, P., Gulick, S. P., Morgan, J. V., and Collins, G. S., "Chicxulub and the Exploration of Large Peak-Ring Impact Craters through Scientific Drilling," *GSA Today* 27, no. 10 (2017).

Chapter 8: The Mirror

1 Lorenz, K., *Studies in Animal and Human Behaviour*, vol. 1 (Harvard University Press, 1970).

2 Davies, N. B., and Brooke, M. D. L., "Cuckoos versus Reed Warblers: Adaptations and Counteradaptations," *Animal Behaviour* 36, no. 1 (1988): 262–84.

3 Patrick, S. C., and Weimerskirch, H., "Personality, Foraging and Fitness Consequences in a Long Lived Seabird," *PloS One* 9, no. 2 (2014): e87269.

4 Brosnan, S. F., and De Waal, F. B., "Monkeys Reject Unequal Pay," *Nature* 425 (2003): 297–99.

5 Johnson, Kristen, "Wernicke's Aphasia," April 30, 2012, YouTube, https://www.youtube.com/watch?v=dWBfgJ-VLOg; Cogmonaut, "Wernicke's Aphasia," January 26, 2010, YouTube, https://www.youtube.com/watch?v=dKTdMV6cOZw.

6 Gallese, V., and Goldman, A., "Mirror Neurons and the Simulation Theory of Mind-Reading," *Trends in Cognitive Sciences* 2 (1998): 493–501.

7 Michael, J., Sandberg, K., Skewes, J., et al., "Continuous Theta-Burst Stimulation Demonstrates a Causal Role of Premotor Homunculus in Action Understanding," *Psychological Science* 25, no. 4 (2014): 963–72.

8 Keysers, C., and Gazzola, V., "Expanding the Mirror: Vicarious Activity for Actions, Emotions, and Sensations," *Current Opinion in Neurobiology* 19, no. 6 (2009): 666–71; Keysers, C., and Gazzola, V., "Hebbian Learning and Predictive Mirror Neurons for Actions, Sensations and Emotions," *Philosophical Transactions of the Royal Society B: Biological Sciences*, 369, no. 1644 (2014): 20130175.

9 Botha-Brink, J., and Modesto, S. P., "A Mixed-Age Classed 'Pelycosaur' Aggregation from South Africa: Earliest Evidence of Parental Care in Amniotes? *Proceedings of the Royal Society B: Biological Sciences* 274, no. 1627 (2007): 2829–34.

10 Jasinoski, S. C., and Abdala, F., "Aggregations and Parental Care in the Early Triassic Basal Cynodonts Galesaurus Planiceps and Thrinaxodon Liorhinus," *PeerJ* 5 (2017): e2875.

11 Hopson, J. A., "Endothermy, Small Size, and the Origin of Mammalian Reproduction," *The American Naturalist* 107 (1973): 446–52; Schmidt-Nielsen, K., *Animal Physiology: Adaptation and Environment* (Cambridge University Press, 1997).

12 Broad, K. D., Curley, J. P., and Keverne, E. B., "Mother-Infant Bonding and the Evolution of Mammalian Social Relationships," *Philosophical Transactions of the Royal Society B: Biological Sciences* 361, no. 1476 (2006): 2199–214.

13 Dumont, G. J., Sweep, F. C. G. J., Van der Steen, R., et al., "Increased Oxytocin Concentrations and Prosocial Feelings in Humans after Ecstasy (3, 4-Methylenedioxymethamphetamine) Administration," *Social Neuroscience* 4, no. 4 (2009): 359–66.

14 Feldman, R., "Oxytocin and Social Affiliation in Humans," *Hormones and Behavior* 61, no. 3 (2012): 380–91.

15 Zak, P. J., Stanton, A. A., and Ahmadi, S., "Oxytocin Increases Generosity in Humans," *PloS One* 2, no. 11 (2007): e1128.

16 Kosfeld, M., Heinrichs, M., Zak, P. J., Fischbacher, U., and Fehr, E., "Oxytocin Increases Trust in Humans," *Nature* 435, no. 7042 (2005): 673–76.

17 Domes, G., Heinrichs, M., Michel, A., Berger, C., and Herpertz, S. C., "Oxytocin Improves 'Mind-Reading' in Humans," *Biological Psychiatry* 61, no. 6 (2007): 731–33.

18 Guastella, A. J., Einfeld, S. L., Gray, K. M., et al., "Intranasal Oxytocin Improves Emotion Recognition for Youth with Autism Spectrum Disorders," *Biological Psychiatry* 67, no. 7 (2010): 692–94.

19 Guastella, A. J., Mitchell, P. B., and Dadds, M. R., "Oxytocin Increases Gaze to the Eye Region of Human Faces," *Biological Psychiatry* 63, no. 1 (2008): 3–5.

20 Seyfarth, R. M., and Cheney, D. L., "Affiliation, Empathy, and the Origins of Theory of Mind," *Proceedings of the National Academy of Sciences* 110, no. S2 (2013): 10349–356.

21 Seed, A. M., Clayton, N. S., and Emery, N. J., "Postconflict Third-Party Affiliation in Rooks, Corvus Frugilegus," *Current Biology* 17, no. 2 (2007): 152–58.

22 Dally, J. M., Emery, N. J., and Clayton, N. S., "Food-Caching Western Scrub-Jays Keep Track of Who Was Watching When," *Science* 312, no. 5780 (2006): 1662–65.

23 Cloutier, S., Newberry, R. C., Honda, K., and Alldredge, J. R., "Cannibalistic Behaviour Spread by Social Learning," *Animal Behaviour* 63, no. 6 (2002): 1153–62.

24 Ciochon, R. L., *Primate Evolution and Human Origins* (Routledge, 2017); Bloch, J. I., and Boyer, D. M., "Grasping Primate Origins," *Science* 298, no. 5598 (2002): 1606–10.

25 Jacobs, G. H., "Evolution of Colour Vision in Mammals," *Philosophical Transactions of the Royal Society B: Biological Sciences* 364, no. 1531 (2009): 2957–67; Hall, M. I., Kamilar, J. M., and Kirk, E. C. "Eye Shape and the Nocturnal Bottleneck of Mammals," *Proceedings of the Royal Society B: Biological Sciences* 279, no. 1749 (2012): 4962–68; Heesy, C. P., and Hall, M. I., "The Nocturnal Bottleneck and the Evolution of Mammalian Vision," *Brain Behavior and Evolution* 75, no. 3 (2010): 195–203.

26 Heesy, C. P., "Seeing in Stereo: The Ecology and Evolution of Primate Binocular Vision and Stereopsis," *Evolutionary Anthropology: Issues, News, and Reviews* 18, no. 1 (2009): 21–35.

27 Carvalho, S., Biro, D., Cunha, E., et al., "Chimpanzee Carrying Behaviour and the Origins of Human Bipedality," *Current Biology* 22, no. 6 (2012): R180–81.

28 Van Schaik, C. P., "Why Are Diurnal Primates Living in Groups?" *Behaviour* 87, nos. 1–2 (1983): 120–44.

29 De Dreu, C. K., Greer, L. L., Van Kleef, G. A., Shalvi, S., and Handgraaf, M. J., "Oxytocin Promotes Human Ethnocentrism," *Proceedings of the National Academy of Sciences* 108, no. 4 (2011): 1262–66.

30 De Dreu, C. K., Greer, L. L., Handgraaf, M. J., et al., "The Neuropeptide Oxytocin Regulates Parochial Altruism in Intergroup Conflict among Humans," *Science* 328, no. 5984 (2010): 1408–11.

31 Dunbar, R. I., "Neocortex Size as a Constraint on Group Size in Primates," *Journal of Human Evolution* 22, no. 6 (1992): 469–93.

Chapter 9: Animals of Abstraction

1 Carew, T. J., Castellucci, V. F., and Kandel, E. R., "An Analysis of Dishabituation and Sensitization of the Gill-Withdrawal Reflex in Aplysia," *International Journal of Neuroscience* 2, no. 2 (1971): 79–98.

2 Kandel, E. R., *Behavioral Biology of Aplysia* (W. H. Freeman and Company, 1979).

3 Woods, S. S., Resnick, L. B., and Groen, G. J., "An Experimental Test of Five Process Models for Subtraction," *Journal of Educational Psychology* 67, no. 1 (1975): 17.

4 Kamii, C., Lewis, B. A., and Kirkland, L. D., "Fluency in Subtraction Compared with Addition," *Journal of Mathematical Behavior* 20, no. 1 (2001): 33–42.

5 Ballard, D. H., *Brain Computation as Hierarchical Abstraction* (MIT Press, 2015).

6 Martin, K. C., Casadio, A., Zhu, H., et al., "Synapse-Specific, Long-Term Facilitation of Aplysia Sensory to Motor Synapses: A Function for Local Protein Synthesis in Memory Storage," *Cell* 91, no. 7 (1997): 927–38; Montarolo, P. G., Goelet, P., Castellucci, V. F., Morgan, J., Kandel, E. R., and Schacher, S., "A Critical Period for

Macromolecular Synthesis in Long-Term Heterosynaptic Facilitation in Aplysia," *Science* 234, no. 4781 (1986): 1249–54.

7 Flexner, J. B., Flexner, L. B., and Stellar, E., "Memory in Mice as Affected by Intracerebral Puromycin," *Science* 141, no. 3575 (1963): 57–59.

8 Castellucci, V. F., Blumenfeld, H., Goelet, P., and Kandel, E. R., "Inhibitor of Protein Synthesis Blocks Longterm Behavioral Sensitization in the Isolated Gill-Withdrawal Reflex of Aplysia," *Journal of Neurobiology* 20, no. 1 (1989): 1–9.

9 Montarolo, P. G., Goelet, P., Castellucci, V. F., Morgan, J., Kandel, E. R., and Schacher, S., "A Critical Period for Macromolecular Synthesis in Long-Term Heterosynaptic Facilitation in Aplysia," *Science* 234, no. 4781 (1986): 1249–54.

10 Caraci, F., Battaglia, G., Bruno, V., et al., "TGF-β1 Pathway as a New Target for Neuroprotection in Alzheimer's Disease," *CNS Neuroscience & Therapeutics* 17, no. 4 (2011): 237–49; Zhang, F., Endo, S., Cleary, L. J., Eskin, A., and Byrne, J. H., "Role of Transforming Growth Factor-β in Long-Term Synaptic Facilitation in Aplysia," *Science* 275, no. 5304 (1997): 1318–20.

11 Kukushkin, N. V., Carney, R. E., Tabassum, T., and Carew, T. J., "The Massed-Spaced Learning Effect in Non-Neural Human Cells," *Nature Communications* 15, no. 1 (2024): 9635.

12 Kukushkin, N. V., Tabassum, T., and Carew, T. J., "Precise Timing of ERK Phosphorylation/Dephosphorylation Determines the Outcome of Trial Repetition during Long-Term Memory Formation," *Proceedings of the National Academy of Sciences* 119, no. 40 (2022): e2210478119.

13 Louie, K., and Wilson, M. A., "Temporally Structured Replay of Awake Hippocampal Ensemble Activity during Rapid Eye Movement Sleep," *Neuron* 29, no. 1 (2001): 145–56.

14 Chalmers, D. J., "Facing up to the Problem of Consciousness," *Journal of Consciousness Studies* 2, no. 3 (1995): 200–219.

15 Clark, A., Friston, K., and Wilkinson, S., "Bayesing Qualia: Consciousness as Inference, not Raw Datum," *Journal of Consciousness Studies* 26, nos. 9–10 (2019): 19–33.

Chapter 10: Fire from Within

1 Herwig, A., and Schneider, W. X., "Predicting Object Features across Saccades: Evidence from Object Recognition and Visual Search," *Journal of Experimental Psychology: General* 143, no. 5 (2014): 1903; Land, M., and Tatler, B., *Looking and Acting: Vision and Eye Movements in Natural Behaviour* (Oxford University Press, 2009).

2 Herwig, A., and Schneider, W. X., "Predicting Object Features across Saccades: Evidence from Object Recognition and Visual Search," *Journal of Experimental Psychology: General* 143, no. 5 (2014): 1903.

3 Weerd, P. D., Gattass, R., Desimone, R., and Ungerleider, L. G., "Responses of Cells in Monkey Visual Cortex during Perceptual Filling-in of an Artificial Scotoma," *Nature* 377, no. 6551 (1995): 731–34.

4 Doss, M. K., Madden, M. B., Gaddis, A., et al., "Models of Psychedelic Drug Action: Modulation of Cortical-Subcortical Circuits," *Brain*, 145, no. 2 (2022): 441–56.

5 Bartolomei, F., Barbeau, E., Gavaret, M., et al., "Cortical Stimulation Study of the Role of Rhinal Cortex in Deja Vu and Reminiscence of Memories," *Neurology* 63, no. 5 (2004): 858–64.

6 Wimmer, K., Nykamp, D. Q., Constantinidis, C., and Compte, A., "Bump Attractor Dynamics in Prefrontal Cortex Explains Behavioral Precision in Spatial Working Memory," *Nature Neuroscience* 17, no. 3 (2014): 431–39.

7 Ackman, J. B., Burbridge, T. J., and Crair, M. C., "Retinal Waves Coordinate Patterned Activity throughout the Developing Visual System," *Nature* 490, no. 7419 (2012): 219–25.

8 Thomson, E. E., Carra, R., and Nicolelis, M. A., "Perceiving Invisible Light through a Somatosensory Cortical Prosthesis," *Nature Communications* 4, no. 1 (2013): 1482.

9 Gindrat, A. D., Chytiris, M., Balerna, M., Rouiller, E. M., and Ghosh, A., "Use-Dependent Cortical Processing from Fingertips in Touchscreen Phone Users," *Current Biology* 25, no. 1 (2015): 109–16.

10 Elbert, T., Pantev, C., Wienbruch, C., Rockstroh, B., and Taub, E., "Increased Cortical Representation of the Fingers of the Left Hand in String Players," *Science* 270, no. 5234 (1995): 305–7.

11 Tononi, G., "An Information Integration Theory of Consciousness," *BMC Neuroscience* 5 (2004): 1–22.

12 Friston, K., "A Theory of Cortical Responses," *Philosophical Transactions of the Royal Society B: Biological Sciences* 360, no. 1456 (2005): 815–36.

13 Lawrence, S. J., Norris, D. G., and De Lange, F. P., "Dissociable Laminar Profiles of Concurrent Bottom-Up and Top-Down Modulation in the Human Visual Cortex," *Elife* 8 (2019): e44422.

14 Shipp, S., "The Importance of Being Agranular: A Comparative Account of Visual and Motor Cortex," *Philosophical Transactions of the Royal Society B: Biological Sciences* 360, no. 1456 (2005): 797–814.

15 García-Cabezas, M. Á., and Barbas, H., "Area 4 Has Layer IV in Adult Primates," *European Journal of Neuroscience* 39, no. 11 (2014): 1824–34.

16 Plato, *Timaeus*, trans. B. Jowett (Echo Library, 2006).

17 Siegel, R. E., "Principles and Contradictions of Galen's Doctrine of Vision," *Sudhoffs Archive* 54, no. 3 (1970): 261–76.

18 Gregg, V. R., Winer, G. A., Cottrell, J. E., Hedman, K. E., and Fournier, J. S., "The Persistence of a Misconception about Vision after Educational Interventions," *Psychonomic Bulletin & Review* 8 (2001): 622–26.

19 Thibodeau, P., "Ancient Optics: Theories and Problems of Vision," in *A Companion to Science, Technology, and Medicine in Ancient Greece and Rome* (John Wiley, 2016), 130–44.

Chapter 11: The Dark Room

1 Friston, K., Thornton, C., and Clark, A., "Free-Energy Minimization and the Dark-Room Problem," *Frontiers in Psychology* 3 (2012): 130; Clark, A., "Whatever Next? Predictive Brains, Situated Agents, and the Future of Cognitive Science," *Behavioral and Brain Sciences* 36, no. 3 (2013): 181–204.

2 Sacks, O., *Awakenings* (Pan Macmillan, 1991).

3 Zhou, Q. Y., and Palmiter, R. D., "Dopamine-Deficient Mice Are Severely Hypoactive, Adipsic, and Aphagic," *Cell* 83, no. 7 (1995): 1197–209.

4 Wardle, M. C., Treadway, M. T., Mayo, L. M., Zald, D. H., and de Wit, H., "Amping Up Effort: Effects of D-Amphetamine on Human Effort-Based Decision-Making," *Journal of Neuroscience* 31, no. 46 (2011): 16597–602.

5 Wyvell, C. L., and Berridge, K. C., "Intra-Accumbens Amphetamine Increases the Conditioned Incentive Salience of Sucrose Reward: Enhancement of Reward 'Wanting' without Enhanced 'Liking' or Response Reinforcement," *Journal of Neuroscience* 20, no. 21 (2000): 8122–30.

6 Otani, S., Daniel, H., Roisin, M. P., and Crepel, F., "Dopaminergic Modulation of Long-Term Synaptic Plasticity in Rat Prefrontal Neurons," *Cerebral Cortex* 13, no. 11 (2003): 1251–56.

7 Sulzer, J., Sitaram, R., Blefari, M. L., et al., "Neurofeedback-Mediated Self-Regulation of the Dopaminergic Midbrain," *NeuroImage* 83 (2013): 817–25.

8 Tik, M., Sladky, R., Luft, C. D. B., et al., "Ultra-High-Field fMRI Insights on Insight: Neural Correlates of the Aha!-Moment," *Human Brain Mapping* 39, no. 8 (2018): 3241–52; Oh, Y., Chesebrough, C., Erickson, B., Zhang, F., and Kounios, J., "An Insight-Related Neural Reward Signal," *NeuroImage* 214 (2020): 116757.

9. Salimpoor, V. N., Benovoy, M., Larcher, K., Dagher, A., and Zatorre, R. J., "Anatomically Distinct Dopamine Release during Anticipation and Experience of Peak Emotion to Music," *Nature Neuroscience* 14, no. 2 (2011): 257–62.

10. Schultz, W., Tremblay, L., and Hollerman, J. R., "Reward Processing in Primate Orbitofrontal Cortex and Basal Ganglia," *Cerebral Cortex* 10, no. 3 (2000): 272–83.

11. Knutson, B., Fong, G. W., Adams, C. M., Varner, J. L., and Hommer, D., "Dissociation of Reward Anticipation and Outcome with Event-Related fMRI," *Neuroreport* 12, no. 17 (2001): 3683–87.

12. Treadway, M. T., Buckholtz, J. W., Cowan, R. L., et al., "Dopaminergic Mechanisms of Individual Differences in Human Effort-Based Decision-Making," *Journal of Neuroscience* 32, no. 18 (2012): 6170–76.

13. Ferster, C. B., and Skinner, B. F., *Schedules of Reinforcement* (Appleton-Century-Crofts, 1957).

14. Wanat, M. J., Kuhnen, C. M., and Phillips, P. E., "Delays Conferred by Escalating Costs Modulate Dopamine Release to Rewards but Not Their Predictors," *Journal of Neuroscience* 30, no. 36 (2010): 12020–27; Fiorillo, C. D., Tobler, P. N., and Schultz, W., "Discrete Coding of Reward Probability and Uncertainty by Dopamine Neurons," *Science* 299, no. 5614 (2003): 1898–1902.

15. Berridge, K. C., and Robinson, T. E., "What Is the Role of Dopamine in Reward: Hedonic Impact, Reward Learning, or Incentive Salience?" *Brain Research Reviews* 28, no. 3 (1998): 309–69.

16. Berridge, K. C., and Kringelbach, M. L., "Pleasure Systems in the Brain," *Neuron* 86, no. 3 (2015): 646–64.

17. Williams, J. T., Christie, M. J., and Manzoni, O., "Cellular and Synaptic Adaptations Mediating Opioid Dependence," *Physiological Reviews* 81, no. 1 (2001): 299–343.

18. Mas-Herrero, E., Ferreri, L., Cardona, G., et al., "The Role of Opioid Transmission in Music-Induced Pleasure," *Annals of the New York Academy of Sciences* 1520, no. 1 (2023): 105–14.

19. Ho, C. Y., and Berridge, K. C., "Excessive Disgust Caused by Brain Lesions or Temporary Inactivations: Mapping Hotspots of the Nucleus Accumbens and Ventral Pallidum," *European Journal of Neuroscience* 40, no. 10 (2014): 3556–72; Williams, J. T., Christie, M. J., and Manzoni, O., "Cellular and Synaptic Adaptations Mediating Opioid Dependence," *Physiological Reviews* 81, no. 1 (2001): 299–343.

20. Damasio, A., Damasio, H., and Tranel, D., "Persistence of Feelings and Sentience after Bilateral Damage of the Insula," *Cerebral Cortex* 23, no. 4 (2013): 833–46.

Chapter 12: In the Beginning Was the Word

1. Heil, M., and Karban, R., "Explaining Evolution of Plant Communication by Airborne Signals," *Trends in Ecology & Evolution* 25, no. 3 (2010): 137–44.

2. Wyatt, T. D., *Pheromones and Animal Behavior: Chemical Signals and Signatures* (Cambridge University Press, 2014).

3. Haddock, S. H., Moline, M. A., and Case, J. F., "Bioluminescence in the Sea," *Annual Review of Marine Science* 2, no. 1 (2010): 443–93; Martini, S., and Haddock, S. H., "Quantification of Bioluminescence from the Surface to the Deep Sea Demonstrates Its Predominance as an Ecological Trait," *Scientific Reports* 7, no. 1 (2017): 45750.

4. Janik, V. M., and Sayigh, L. S., "Communication in Bottlenose Dolphins: 50 Years of Signature Whistle Research," *Journal of Comparative Physiology A* 199 (2013): 479–89.

5. Seyfarth, R. M., Cheney, D. L., and Marler, P., "Vervet Monkey Alarm Calls: Semantic Communication in a Free-Ranging Primate," *Animal Behaviour* 28, no. 4 (1980): 1070–94.

6. Humboldt, W. von, *Linguistic Variability & Intellectual Development* (University of Miami Press, 1971).

7. Everett, D., *Don't Sleep, There Are Snakes: Life and Language in the Amazonian Jungle* (Profile Books, 2010).

8. Colapinto, J., "Has a Remote Amazonian Tribe Upended Our Understanding of Language?" *New Yorker*, April 16, 2007.

9. Everett, D., "Cultural Constraints on Grammar and Cognition in Pirahã: Another Look at the Design Features of Human Language," *Current Anthropology* 46, no. 4 (2005): 621–46; Everett, D. L., "Challenging Chomskyan Linguistics: The Case of Pirahã," *Human Development* 50, no. 6 (2007): 297–99.

10. Sakel, J., "Acquiring Complexity: The Portuguese of Some Pirahã Men," *Linguistic Discovery* 10, no. 1 (2012).

11. Botha, R., "On Homesign Systems as a Potential Window on Language Evolution," *Language & Communication* 27, no. 1 (2007): 41–53.

12. Pyers, J. E., and Senghas, A., "Language Promotes False-Belief Understanding: Evidence from Learners of a New Sign Language," *Psychological Science* 20, no. 7 (2009): 805–12; Pyers, J. E., Shusterman, A., Senghas, A., Spelke, E. S., and Emmorey, K., "Evidence from an Emerging Sign Language Reveals That Language Supports Spatial Cognition," *Proceedings of the National Academy of Sciences* 107, no. 27 (2010): 12116–20; Senghas, A., "Intergenerational Influence and Ontogenetic Development in the Emergence of Spatial Grammar in Nicaraguan Sign Language," *Cognitive Development* 18, no. 4 (2003): 511–31.

13 Pyers, J. E., Shusterman, A., Senghas, A., Spelke, E. S., and Emmorey, K., "Evidence from an Emerging Sign Language Reveals That Language Supports Spatial Cognition," *Proceedings of the National Academy of Sciences* 107, no. 27 (2010): 12116–20.

14 Gordon, P., "Numerical Cognition without Words: Evidence from Amazonia," *Science* 306, no. 5695 (2004): 496–99.

15 Berlin, B., and Kay, P., *Basic Color Terms: Their Universality and Evolution* (University of California Press, 1991).

16 Malotki, E., *Hopi Time: A Linguistic Analysis of the Temporal Concepts in the Hopi Language* (Walter de Gruyter, 2011).

17 Senghas, A., "Intergenerational Influence and Ontogenetic Development in the Emergence of Spatial Grammar in Nicaraguan Sign Language," *Cognitive Development* 18, no. 4 (2003): 511–31.

18 Rensberger, B., "Chimpanzees Teach Sign Language: Scientists Told That Apes in Experiments Instruct Each Other," *Washington Post*, May 29, 1985.

19 Fouts, R. S., and Fouts, D. H., "Loulis in Conversation with the Cross-Fostered Chimpanzees," in *Teaching Sign Language to Chimpanzees*, ed. R. A. Gardner, B. T. Gardner, and T. E. Van Cantfort (State University of New York Press, 1989), 293–307.

20 Deacon, T. W., *The Symbolic Species: The Co-evolution of Language and the Brain* (W. W. Norton, 1998).

Index

abiogenesis, 22
abstract art, 197–98
abstraction: brain and, 164–68, 173, 241; cerebral cortex and, 196–99; computers and, 171, 173, 182; cortical columns and, 196–201, 204; human brain and, 166–68, 173, 182, 183, 250; human cognition and, 169; linguistic recursion and, 240–41; linguistic trees of, 241–42; memory and, 173–74, 178; neuronal spike and, 173, 175; neurons and, 169, 171, 177; pattern and, 176, 250; sea slugs and, 165–69, 179, 182–83; tree of, 199–202, 204, 241–42, 250
actin, 89
action: as embodied hallucination, 204; top-down information flow and, 202–3
action potential. *See* neuronal spike
Akhmatova, Anna, 181
albatross, 143–44
algae, 99
alleles, 60
All Yesterdays (Conway and Kosemen), 122
amino acids, 13, 14, 16–17, 22; molecules and, 11
amnion, 114–16
amniotes, 115–19, 147
amniotic eggs, 147
amphibians: fish and, 114–16; heart of, 140, 141; of Paleozoic era, 114–16; water and, 114–16
anabolism, 4
animal cells, 80, 95; cells walls lacking in, 79; multicellularity, 81–85
animals: body temperature of, 131–33; brains of, 185; breathing of, 77, 78; of Cambrian explosion, 93–94, 95; digestion and, 77, 85–87, 117–19; dinosaurs and, 113–14; of Ediacaran period, 93; essence of kingdom of, 77; evolution of, 86, 94; food and, 78, 82, 84–87; as herbivores, 117–18; Mesozoic era and, 113–14; motion of, 78–82, 84–85, 89–90, 94, 95–96; muscles of, 89; plants and, 80–81, 102, 103, 117–18; sponge larvae and, 84–87; warm-bloodedness of, 132–33, 136. *See also specific animals*
anisogamy, 66, 67
antibiotics, 53–54
ants, 67–69
Aplysia sea slugs. *See* sea slugs
archaea, 41–43, 46; Asgard, 47–48, 51; bacteria and, 51–52; Loki, 47–48, 51
Arrival (film), 90
art, abstract, 197–98
arthropods: exoskeletons of, 104–6, 110; vertebrates compared to, 99, 105–6
Asgard archaea, 47–48, 51
associative cortex, 198–99
asteroid impacts, 137–38
atoms, 5, 7, 9, 13, 22, 71, 72
ATP synthase, 134–35
autoshaping, 220

bacteria, 41–43, 49–50; antibiotic resistant, 53–54; archaea and, 51–52; breathing of, 47, 51, 52, 77; choanoflagellates and, 82–83; as collectivists, 71; cyanobacteria, 47, 52, 99; DNA of, 60, 63; eukaryotes and, 52–53, 54; as food, 82–83, 85, 91; as formidable, 53–54; genes of, 60, 70; germ theory and, 53; multicellularity and, 70; photosynthesis and, 76; reproduction of,

277

60, 63–65, 67; sexual reproduction and, 63–65, 70; sponges and, 82–83; strain, 70
Barto, Agnia, 161
basal ganglia, *213*, 215, 216
Beagle. See HMS *Beagle*
behaviorism, 177
bilateral symmetric bodies, 91
bilaterians, 91–92
binocular rivalry, *184*, 184–85, 191
biodiversity, 35–36, 38, 250
biology: evolution and, 17, 34–35, 41; Orgel's Second Rule, 57–58
birds: dinosaurs and, 123, 128; eggs of, 148; lungs of, 126–28; mammals and, 140–41, 144, 150, 152; parenting of, 144
birthing, by mammals, 147, 149
Black Square (painting), 197–98
bodies: bilateral symmetric, 91; radially symmetric, 90–91
body temperature: of animals, 131–33; on land, 131–32; of lizards, 133; of mammals, 133–36; of synapsids, 132–33, 147; warm-bloodedness and heat threshold of, 136; in water, 131
Book of Optics (Ibn al-Haytham), 205–6
bottom-up information flow: computers and, 185; through cortical columns, 199–204; human brain and, 183–85, *197*; human perception and, 188–89, 195–96
brain: abstraction and, 164–68, 173, 241; ancient philosophers on, 207; basal ganglia, 215, 216; computers and, 173, 181, 182; consciousness and, viii, 191–92, 202; creation of meaning by, 177; dopamine and, 212–16, *213*; human perception and animal, 185; memory and, 173–74; neuronal spike of, 161–62, 169–71; oxytocin and reflection by, 149; recursion and, 240, 241; substantia nigra, 212; usefulness of, 209; vision and, 207. *See also* human brain

brain imaging, 177
brain science, on mirroring, 146–47
breathing: of animals, 77, 78; of bacteria, 47, 51, 52, 77; of humans, 52, 56, 78; of insects, 108–10; oxygen and, 108–10, 126–28; passive and active, 108–9; of plants, 108; of sauropsids, 125–26, 128–29; tracheal, 108, 110; unidirectional, 126–29; of vertebrates, 108–9
Brodsky, Joseph, 113
Bryullov, Karl, 120, 122
Buddha, 224, 226, *226*
Buddhism, vii, 226–27
Bunin, Ivan, 229

Cambrian explosion, 93–94, 95, 98, 103, 105
capuchin monkeys, 144–45
carbon, 5, 7, 9, 22
carbon dioxide: photosynthesis and, 44–45, 102–3; plants and, 102–3
Carboniferous dragonflies, 103–4, 106, 107–9
Carboniferous period, 103–4, 106, 107–9; oxygen levels during, 124–25
catabolism, 4
cell membranes, 49, 95; moving, 48, 51
cells, 71; complexity of, 59; eukaryotic, 49–50, 58–59; muscle, 89; nutrient energy harvesting by, 134–35; place, 177; plant, 76, 79, 80; sex, 61–62, 65–67, 70; single, 81, 82, 84, 89, 95. *See also* multicellular organisms, multicellularity
cellulases (cellulose-digesting enzymes), 117–18
cell walls, 79
cenotes, 137–38
Cenozoic era, 138–39, 154
centers of creation, 28, 29
central dogma, 14, 16, 17, 22
cerebral cortex: abstraction and, 196–99; associative cortex, 198–99; consciousness

and, 202–4; cortical columns of, 191–204; dark room problem and, 210, 213, 218; desire and, 225–26; dopamine and, 212, 216, 218–19; electrical stimulation and mapping of, 190–91, *191*, 195–96; fusiform face area, 189–90; group life and, 155, 157; happiness and, 227; hearing and, 198; hippocampus, 195; human perception and, 189, 192–93; human senses and, 196–98, *199*; layers of, 203–4; mammalian senses and, 131; map of, 190–91, *191*, 196, *199*; motion and, 204; motivations and, 225; motor cortex, 203–4; neuronal connections in, 194; of primates, 155, 157; reality alignment of, 210, 218; relational map of world in, 195; reward system and, 210, 221, 223–26, 227; sensory cortex, 203–4; signal flow through, *197*, 204; visual cortex, 196–98, 204
Chalmers, David, 178
Chekhov, Anton, 97
Chiang, Ted, 90
Chicxulub impact, 137–38
chloroplasts, 52
choanoflagellates, 82–84
Chomsky, Noam, 233–36, 237–40, 242
chromosomes: of eukaryotes, 60; human genes and, 60–61; human reproduction and, 60–62; mixing genes and, 61; sexes and, 65; X and Y, 65
cilia, 81–82, 89, 90
ciliates, 81
Clark, Andy, 178–79
cnidarians, 87–89
cobra, spectacled, 31–33
coevolution, of human brain and human language, *244*, 244–45
cognition, 178; human, 162, 169, 239; neurons and, 162, 169
cold, nocturnality and, 129, 130, 131

cold-bloodedness, 133
complementarity, 12
complexity: of cells, 59; of eukaryotes, 54–56, 59, 67, 72, 93–94; of human brain, 180; of human language, 244; of humans, 40–41, 43, 55–56; perfection and, 55–56; primordial ocean and, 41
computers: abstraction and, 171, 173, 182; bottom-up information flow and, 185; brain and, 173, 181, 182; human brain and, 185; top-down information flow and, 185
consciousness: brain and, viii, 191–92, 202; cerebral cortex and, 202–4; cortical columns and, 191–92, 202–4; first-personness and, 178–79; hierarchical predictive coding theory on, 202–4; IIT on, 202, 204; nature of, viii, 178
convergent evolution, 140, 162
Conway, John, 122
cortex. *See* cerebral cortex
cortical columns, 195; abstraction and, 196–201, 204; bottom-up information flow through, 199–204; consciousness and, 191–92, 202–4; human perception and, 192–93; human senses and, 196–98; neurons of, 192–93; synaptic plasticity and, 193–94; top-down information flow through, 199–204; tree of abstractions and, 199–202, 204
Crick, Francis, 9
crocodiles, 128
ctenophores, 81–82
culture, 246–47, 249
cyanobacteria, 47, 52, 99
cynodonts, 124; mammals and, 121, 130–31; nocturnality of, 130–31

dark room problem, 210, 213, 218
Darwin, Charles, 20; *The Origin of Species*, 25–26, 33, 37; theory of evolution, 25–34, 37–38

Index **279**

DarwinTunes, 33
Dawkins, Richard, 38
deaf people, 236–40, 242–43
deinonychus, 122–23
dendrites, 169–70
desire: dopamine and, 227; sources for, 225–26; suffering, pleasure and, 224, 226
de Waal, Frans, 144–45
digestion: animals and, 77, 85–87, 117–19; cellulases and, 117–18; fungi and, 76–77, 118, 125; jellyfish and, 87, 88, 92; sponges and, 85–87, 118; of wood, 125
dinosaurs: animals and, 113–14; birds and, 123, 128; Cenozoic era and, 138, 139; extinction of nonavian, 123; fossils of, 122; humans and, 113; ideas about, 122–23; mammals and, 136–37, 138, 141, 150; mass extinction of, 138; Mesozoic era and, 113, 121, 123–24, 128–30, 152
dispersal: plants and, 80–81, 101; sponges and, 84–85
diurnality: of humans, 129–30; of primates, 152, 154
diversification, sensory, 131
diversity: evolution, unity and, 36, 250. *See also* biodiversity
DNA, 11, 13–14, 16–18; of bacteria, 60, 63; double helix, 9, 12; sexual reproduction and replication of, 60, 63–64; "we" and, 245
Dobzhansky, Theodosius, 34–35, 36
dolphins, 129, 150
dopamine, 210, *211*; brain and, 212–16, *213*; cerebral cortex and, 212, 216, 218–19; desire and, 227; as "do more of that" chemical, 214–18; economics of, 227; evolution and, 223; expectations, surprises and, 218–19; as "figure this out" chemical, 218, 219–20; memory and, 214–15; motivation and, 222–23, 227; neurons, *223*; pattern seeking, expectation and, 220; pleasure and, 214, 218–19, 221–22; thinking and, 227; unexpected success and, 216–17, *217*, 222–23
dragonflies: Carboniferous, 107–9; giant Paleozoic, 103–4, 106, 107–9; larvae, 97–98
"The Dress" (photo), 186

earthworm, 91–92
Ediacaran period, 96; fossils of, 93
eggs: amniotic, 147; of birds, 148; external and internal, 148; of mammals, 148; sex cells, 66–67, 70; of synapsids, 147–48
embryo development, 86; evolution and, 95
emergence, 71–72
encephalitis lethargica, 210–12, 219
endocannabinoids, 221
endoskeletons, of vertebrates, 104, 105–6
energy: cellular harvesting of nutrient, 134–35; heat production and waste of, 135; humans and nutritional, 136; mammal consumption and expenditure of, 135–36
enzymes, 85–86, 117–18
epilepsy, 189–90, 192
epithelia, 95
escape velocity, *243*, 243–44
essences: ideas compared to, 179, 250; ideas of nature as, 4, 32–33, 179, 249; of kingdom of animals, 77; of life, 23; of nonlife, 23; selection and, 32–33
eukaryotes, 46, 47–48, 89; bacteria and, 52–53, 54; cells of, 49–50, 58–59; chromosomes of, 60; complexity of, 54–56, 59, 67, 72, 93–94; emergence and, 72; evolution of, 58, 59, 93–94, 99–100; food and, 88; genes of, 60; humans and, 41–43, 55–56, 75, 157; hypercells and, 71; individuality and, 52, 62, 64, 71; kingdoms of, 75–77, 81; land and,

99–100; as multicellular organisms, 59–60, 67, 82; origins of, 51–53, 86; sexual reproduction of, 59–60, 66–67; survival pursuit by, 75
eukaryotic genomes, 60
eusociality, 68
Everett, Daniel, 234–35, 236, 240, 242
evolution: of animals, 86, 94; biodiversity and, 35–36, 38, 250; biology and, 17, 34–35, 41; child understanding of, 250; coevolution of human brain and human language, *244*, 244–45; convergent, 140, 162; Darwin's theory of, 25–34, 37–38; directed, 37; diversity, unity and, 36, 250; dopamine and, 223; embryonic development, 95; emergence and, 72; of eukaryotes, 58, 59, 93–94, 99–100; humans and, 57–58, 72, 156–57, 162; of insects, 98, 104, 109–11, 244–45; intelligent design and, 25–29, 34–36, 38; land and, 98–101; of mammals, 150; nature's ideas and, 32–33; Orgel's Second Rule and, 57–58; Oxford evolution debate, 25–27; of plants, 99–103, 244–45; religion and, 25–28, 36, 37, 39–40; tree of life of, ix, 38, 39–42
evolutionary biologists, 17, 35
evolutionary biology, 41
exoskeletons: of arthropods, 104–6, 110; of insects, 104–6, 110
extinctions, mass: Chicxulub impact and, 137–38; Great Dying, 120–21, 123–25, 130, 132, 138
extramission theory, 205–7
eyes, 152–53, 205–7

face recognition, 189–90
fat storage, 135
finches, Galapagos, 27–30, 35
first-personness, 178–79; language and, 229
fish, 129, 147; amphibians and, 114–16; gills of, 125, 126; land and, 114
Fitzroy, Robert, 26, 28, 29
flagella, 81, 82, 90
food: animals and, 78, 82, 84–87; bacteria as, 82–83, 85, 91; eukaryotes and, 88; fungi and, 78; gut and, 91–94; polyps and, 91–92; sponges and, 85–87; worms and, 91–94
fossils: of dinosaurs, 122; of Ediacaran period, 93
Friston, Karl, 178–79
fungal roots (mycorrhiza), 100
fungi, 78, 99, 100; digestion and, 76–77, 118, 125

Galapagos Islands, 27–30, 35
genes, 9, 11, 13–14, 16–18; of bacteria, 60, 70; of eukaryotes, 60; of humans, 60–61, 70. *See also* DNA; RNA
genetic sequencing, 41
germ line, 69–71, *246*
germ theory, 53
glutamate, 215
Granovich, Andrei, 71–72
Great Dying (end-Permian extinction), 120–21, 123–25, 130, 132, 138
group defense: mammalian social instinct and, 154; primates and, 154, 156
group life: cerebral cortex and, 155, 157; human intelligence and, 154, 155–57; oxytocin and, 154; of primates, 154, 156
gut, 91–94, 96

Haeckel, Ernst, 39–40, 41
hallucination: action as embodied, 204; human perception as, 186, 188, 190, 203–4; top-down information flow and, 203–4
happiness, cerebral cortex and, 227
Hawkins, B. W., 122

hearing: cerebral cortex and, 198; of mammals, 131, 140
heart: of amphibians, 140, 141; of mammals, 140, 141; of sauropsids, 140; of synapsids, 140
herbivores, 117–18
heron, 98
hierarchical predictive coding theory, 202–4
hierarchical predictive processing, 178
hippocampus, 11, 195
HMS *Beagle*, 26, 27, 29
Hodgkin, Alan, 173
Homo sapiens, 43
Hooker, Joseph Dalton, 25, 26
human brain, 55–56; abstraction and, 166–68, 173, 182, 183, 250; bottom-up information flow and, 183–85, *197*; coevolution of human language and, *244*, 244–45; complexity of, 180; computers and, 185; human language and, 233, *244*, 244–45; human misalignment with, 209; human perception and, 207; memory and, 173–74; mental folders, 167–68; midbrain, 212; multicellularity and, 71; neurons of, 161–62, 174; social brain hypothesis, 156–57; top-down information flow and, 183–85, *197*, 207; walnut and, 189
human cognition: abstraction and, 169; human language and, 239; neuronal spike and, 162; neurons and, 169
human experience, meaning of, ix, 4
human genes, 60–61, 70
human genomes, 60
human intelligence, 154–57
human language, 157; Chomsky linguistics, 233–36, 237–40; coevolution of human brain and, *244*, 244–45; as cognitive virus, 243–44; complexity of, 244; escape velocity of, 243–44; as externalized or internalized cognition, 239; first-personness and, 229; generative power of, 230; human brain and, 233, *244*, 244–45; human cognition shaped by, 239; learning, 233; meaning in, 230–33; of Pirahã, 234–36, 239–40, 242; primates learning, 242; recursion of, 230–36, 240–42; words of, 229–30, 231–33
human life, 156–57, 209–10
human perception, 182; animal brains and, 185; bottom-up information flow and, 188–89, 195–96; cerebral cortex and, 189, 192–93; cortical columns and, 192–93; as hallucination, 186, 188, 190, 203–4; human brain and, 207; of music, 221–22; as objective, 186; peripheral vision and, 187, *187*; reality as illusion in, 186; top-down information flow and, 185, 189, 195–96; vision, *184*, 184–88, *187*
human reproduction: choice in, 64–65; chromosomes and, 60–62; individuality and, 60, 64, 71; sex cells and, 61–62
humans: as ancestors of synapsids, 147; breathing of, 52, 56, 78; complexity of, 40–41, 43, 55–56; differences between primates and, 150; dinosaurs and, 113; as diurnal, 129–30; emergence and, 72; eukaryotes and, 41–43, 55–56, 75, 157; evolution and, 57–58, 72, 156–57, 162; germ line in, 70–71, *246*; Homo sapiens, 43; irony of existence of, 226; memory of, 174; mirroring by, 145–47; nutritional energy of, 136; oxygen and, 56; oxytocin and positive interaction between, 149; as perfect species, 39–40, 55–56; Precambrian worms and, 162, 169; sea slugs and, 162, 164, 166–67, 169, 174, 180; size of, 110; soma in, 70–71; specialness of, 249, 251; suffering of, 223–26; as supreme species, 39, 40, 43; synapsids as ancestors of, 132; uncertainty obsession of, 219, 220; vision of, 152–53, 184

human senses, 196–98, *199*
Huxley, Andrew, 173
Huxley, Thomas, 26
hydra, 87
hydrodynamics, 83–84
hydrogen sulfide, 45
hydrothermal vents, 20–22
hypercells, 71. *See also* multicellular organisms, multicellularity

Ibn al-Haytham, 205–6, *206*
ichthyosaurs, 129
ideas, viii; essences compared to, 179, 250; ideas of nature and human, 249; reality basis of, 3–4; sea slugs and, 179; selection and, 34
ideas of nature. *See* nature, ideas of
IIT. *See* integrated information theory
illusion: reality as, 186; visual, 181, *181*, 182, 183, 184
incest, 64
Indian snake charmers, 31
individuality: eukaryotes and, 52, 62, 64, 71; human reproduction and, 60, 64, 71; legacy and, 247; life of, 245–47; multicellular organisms and, 225; proteins and "I," 245; "we" and "I," 245–47
information; top-down information flow: IIT, 202, 204; we-life as life of, 245. *See also* bottom-up information flow
inheritance, 29–30, 33
insects: breathing of, 108–10; coevolution of flowering plants and pollinating, 244–45; evolution of, 98, 104, 109–11, 244–45; exoskeletons of, 104–6, 110; invincibility of, 111; on land, 98–99, 104–6, 109; of Paleozoic era, 109–10; size of, 110–11; vertebrates and, 98, 104, 105–6, 110–11, 114, 139. *See also specific insects*

integrated information theory (IIT), 202, 204
intelligence, human, 154–57
intelligent design, 25–29, 34–36, 38
invertebrates, 104–5
invertebrate zoology, 107
isogamy, 66

jellyfish, 87–88, 89–92, 96

Kandel, Eric, 174
Kepler, Johannes, 206
Kharms, Daniil, 39
koan, vii–viii, ix, 31, 227
Kosemen, C. M., 122

land: body temperature on, 131–32; eukaryotes and, 99–100; evolution and, 98–101; fish and, 114; insects on, 98–99, 104–6, 109; lichen and, 99–100; plants and, 99–103, 116; vertebrates on, 98, 104, 105, 118, 141
language: culture and, 246–47; NSL, 237–40, 242; primates learning sign, 242; protolanguages, 230. *See also* human language
The Last Day of Pompeii (painting), 120, 122
The Last of Us (tv series), 77
L-DOPA, 211–12
learning, pattern finding and, 176
Lees, Abigail, 143
Leibniz, Gottfried, 176–77, 178
Lermontov, Mikhail, 209
lichens, 99–100
life: as continued lineage, 245, 247; essences of, 23; group, 154, 155–57; human, 156–57, 209–10; individual, 245–47; life of information as we-, 245; matter and, 245; nonlife and, 22, 23, 30; origin of, ix, 12, 18, 20–23;

photosynthesis and, 44; social, 155–57; tree of, ix, 38, 39–42
life, domains of, 41–43. *See also* archaea; bacteria; eukaryotes
light: photosynthesis and, 44–45, 102; plants and, 102–3
listrosaurs, 121, 124
lizards, body temperature of, 133
Loki archaea, 47–48, 51
Lost City hydrothermal field, 21–22
lungs, 125–29

macrophage, 49
Malevich, Kazimir, 197–98
mammalian senses: cerebral cortex and, 131; hearing, 131, 140; sensory diversification, 131; vision, 130–31, 152. *See also* human senses
mammalian social instinct, 141, 144, 146, 147; group defense and, 154; maternal reflection and, 150
mammals: birds and, 140–41, 144, 150, 152; birth of, 147, 149; body temperature of, 133–36; Cenozoic era and, 138–39; Chicxulub impact and, 138; cynodonts and, 121, 130–31; dinosaurs and, 136–37, 138, 141, 150; eggs of, 148; energy consumption and expenditure of, 135–36; evolution of, 150; heart of, 140, 141; maternal care by, 148; mirroring by, 145–47, 156; nocturnality of, 130–31; parenting of, 144; speed of, 136, 183; vision of, 130–31, 152; warm-bloodedness of, 133–36, 140. *See also specific mammals*
materialism, science and, viii
maternal care, 147–49
maternal reflection, 150
Mayakovski, Vladimir, 25
meaning: brain creation of, 177; of human experience, ix, 4; in human language, 230–33; of natural processes, viii–ix

meditation, 226
medusas, 87–88, 89. *See also* jellyfish
meme, 246–47
memory: abstraction and, 173–74, 178; brain and, 173–74; dopamine and, 214–15; episodic, 195; hippocampus and, 195; of humans, 173–74; muscle, 215; neurons and, 176, 177; of sea slugs, 174–75, 179; synaptic plasticity and, 175; working, 193
mental: human brain mental folders, 167–68; synaptic plasticity, physical and, 179–80; wall between physical and, 176–79
Mesozoic era: animals in, 113–14; dinosaurs in, 113, 121, 123–24, 128–30, 152; end of, 137–38; oxygen decrease during, 124; sauropsids of, 121, 123–25, 128–29; synapsids of, 121, 124–26, 129–30, 133, 140, 147
metabolism, 4–5, 7, 22
midbrain, 212
Miller, Stanley, 20
Miller-Urey experiment, 20
mind, 178–79; ancient philosophers on, 205–7; nature and, 250, 251; as subjective, 185
"*Minesweeper*" experiments, 190, *191*, 192, 195
mirroring: brain science on, 146–47; by humans, 145–47; by mammals, 145–47, 156; by primates, 144–45, 147
mirror neurons, 146
mitochondria, 52, 134–35
molecules, 3, 4, 7, 71, 72; proteins; RNA; *specific molecules*; amino acids and, 11; organic, 5; origin of life and, 12, 20, 22; ribosomes and, 16. *See also* DNA
mollusks, 174. *See also* sea slugs
monkeys: capuchin, 144–45; leaping by, 151
motion: of animals, 78–82, 84–85, 89–90, 94, 95–96; of bilaterians, 91–92; cerebral

cortex and, 204; of jellyfish, 90–91; multicellular organisms and, 82, 90; plants and, 79–81; of single cells, 81, 82, 84, 90

motivations: balance between reward and, 227; dopamine and, 222–23, 227; misalignment with, 222–23; sources of, 225; survival and, 223

motor cortex, 203–4

mouth, 100

multicellular organisms, multicellularity: animals as, 81–85; bacteria and, 70; epithelia of, 95; eukaryotes as, 59–60, 67, 82; germ line in, 69–71; human brain and, 71; individuality and, 225; motion of, 82, 90; plants and, 76; single cells and, 95; soma in, 69–71; sponges as, 82–85

muscle cells, 89

muscle memory, 215

muscles, of animals, 89

mushrooms, abstract ideas and, 166, 167

music, human perception of, 221–22

mycorrhiza (fungal roots), 100

myosin, 89

"My Wife and My Mother-in-Law" (visual illusion), 181, *181*, 182, 183, 184

Nabokov, Vladimir, 143

naloxone, 221

natural processes, meaning of, viii–ix

natural selection, 27, 29, 33, 72

nature: emergence in, 72; human experience and, ix; mind and, 250, 251

nature, ideas of: as essences, 4, 32–33, 179, 249; evolution and, 33; human ideas and, 249; human mind and, 250; meaning of human experience and, 4; science and, 4

nemertine worms, 107–8

nerves, 90

nervous system, 89, 90

neuronal spike (action potential), 161, 169–71, *172*; abstraction and, 173, 175; human cognition and, 162; physical, mental and, 179–80

neurons, 89, *170*; abstraction and, 169, 171, 177; cerebral cortex and connections between, 194; cognition and, 162, 169; of cortical columns, 192–93; dendrites of, 169–70; dopamine, *223*; excitatory, 192, 193; of human brain, 161–62, 174; inhibitory, 192, 193; memory and, 176, 177; mirror, 146; of sea slugs, 164–66, 170, 175; synapses of, 169–71, 174–76

neuroscience, on vision, 207

neuroscientists, 177, 178

neurotransmitters, 169–72, 210

Nicaraguan school for deaf, 236–40, 242, 243

Nicaraguan Sign Language (NSL), 237–40, 242

nocturnal bottleneck hypothesis, 130, 136

nocturnality: cold and, 129, 130, 131; of cynodonts, 130–31; of mammals, 130–31; of primates, 152, 154; of synapsids, 130–33, 136; vision and, 130–31, 152

nonlife, 22, 23, 30

Nordic mythology, 47–48

NSL. *See* Nicaraguan Sign Language

nucleic acids, 14, 16

nucleotides, 12–14, 16, 20, 21, 22

objective: human perception as, 186; subjective and, 178–80

ocean, primordial, 41

octopus, 104

one hand clapping koan, vii–viii, ix

opioids, 221

organisms, superorganisms, 69, 71. *See also* multicellular organisms, multicellularity

Orgel, Leslie, 58

Orgel's First Rule, 58

Orgel's Second Rule, 57–58

origin of life: hydrothermal vents and, 20–22; meaning of human experience and, ix; molecules and, 12, 20, 22; nucleotides and, 20; RNA and, 18, 20, 21; theories of, 20–23

The Origin of Species (Darwin), 25–26, 33, 37

Ostrom, John, 122–23

Oxford evolution debate, 25–27

oxygen, 7, 9, 22; breathing and, 108–10, 126–28; Carboniferous period levels of, 124–25; humans and, 56; lungs and, 126; Mesozoic era and, 124; origin of eukaryotes and, 51–52; Permian period levels of, 124–25, 129; photosynthesis and, 44–47; plants and, 103; respiration and, 46–47, 51–52

Oxygen Holocaust, 46, 51, 52

oxytocin, 148–49, 154

Paleozoic era, 98; amniotes of, 115–19, 147; amphibians of, 114–16; giant Carboniferous dragonflies in, 103–4, 106, 107–9; insects of, 109–10. *See also specific periods of Paleozoic*

parenting, 144, 148–49

Parkinson's disease, 210–12

pattern: abstraction of, 250; finding, 176; resolution, 222; seeking, 220

Penfield, Wilder, 190, *191*, 195

penicillin, 53

perception. *See* human perception

perfection: complexity and, 55–56; human species as, 39–40, 55–56

peripheral vision, 187, *187*

Permian period, 120–21, 124–25, 129–30

philosophers, ancient, 4; on brain, 207; on human mind, 205–7

photosynthesis, 43, 52; bacteria and, 76; carbon dioxide and, 44–45, 102–3; life and, 44; light and, 44–45, 102; oxygen and, 44–47; of plants, 76, 79, 102–3; water and, 45–46, 102

physical: synaptic plasticity, mental and, 179–80; wall between mental and, 176–79

pigeon experiments, 220

Pirahã (tribe), 234–36, 239–40, 242

place cells, 177

plant cells, 76, 79, 80

plants: amniotes and, 116; animals and, 80–81, 102, 103, 117–18; breathing of, 108; carbon dioxide and, 102–3; coevolution of insect pollinators and flowering, 244–45; communication between, 230; dispersal and, 80–81, 101; evolution of, 99–103, 244–45; flowering, 80–81; fungi and, 100; land and, 99–103, 116; light and, 102–3; motion and, 79–81; multicellularity of, 76; origin of kingdom of, 52; oxygen and, 103; photosynthesis of, 76, 79, 102–3; stems of, 101–2, 125; vertebrates and, 117–18; water and, 100–103, 116

Plato, 4, 205–7, *206*

pleasure: dopamine and, 214, 218–19, 221–22; "liking chemicals" and, 221; pattern resolution as, 222; puzzling source of, 222; suffering, desire and, 224, 226

pollen, 101

polyps, 87–88, 89–92

Pompeii, 120

Precambrian worms, 162, 169

primates: cerebral cortex of, 155, 157; change from nocturnality to diurnality, 152, 154; differences between humans and, 150; group defense and, 154, 156; group life of, 154, 156; leaping by, 151–52; mirroring by, 144–45, 147; monkeys, 144–45, 151; oxytocin and, 154; sign language learning by, 243; tree life of, 150–51, 153–54; vision of, 152

primatology, 144–45
prokaryotes, 41–42, 55
Prometheus archaea, 48, 51
proteins, 9, 11–14, 16–18; cell membranes and, 49, 51, 95; "I" individuality and, 245; UCP1, 134–35
protolanguages, 230
Pushkin, Alexander, 75
Pythagoras, 27

radially symmetric bodies, 90–91
rats, mirroring by, 145–46
reality: cerebral cortex and, 210, 218; ideas based in, 3–4; as illusion in human perception, 186; imagination and, 249, 251
recursion: abstraction and linguistic, 240–41, 242; brain and, 240, 242; linguistic, 230–36, 240–42; NSL and, 240, 242
religion: evolution and, 25–28, 36, 37, 39–40; tree of life and, 39–40
reproduction; sexual reproduction: of bacteria, 60, 63–65, 67; oxytocin and, 149. *See also* human reproduction
respiration, 46–47, 51–52. *See also* breathing
reward, balance between motivation and, 227
reward system, 210, 221, 223–26, 227
ribosome, 16, 17
RNA, 14, 16–17, 18, 20, 21
roundworms, 110
Russian universities, 106–8

Sacks, Oliver, 211–12
salamander, 97, 98
Sapir-Whorf hypothesis, 239
sauropsids: breathing of, 125–26, 128–29; in Cenozoic era, 139; convergent evolution of, 140; heart of, 140; of Mesozoic era, 121, 123–25, 128–29; running by, 129, 140; skulls of, 140; synapsids compared to, 114, 116–17, 119–21, 125–26, 129, 138, 139–40
science: ideas of nature and, 4; materialism and, viii; neuroscience, 177, 178, 207; one hand clapping koan and, vii–viii; tree of life and, 41; on wall between physical and mental, 176–79
sea slugs: abstraction and, 165–69, 179, 182–83; deconstruction of, 162–63; humans and, 162, 164, 166–67, 169, 174, 180; ideas and, 179; memory of, 174–75, 179; neurons of, 164–66, 170, 175; siphon of, 164–66, 175
selected breeding, 33
selection, 30; essences and, 32–33; ideas and, 34; natural, 27, 29, 33, 72
senses: human, 196–98, *199*; mammalian, 130–31, 140, 152
sensory cortex, 203–4
sensory diversification, 131
Seven Worlds, One Planet (TV series), 143
sex cells: anisogamy and, 66, 67; eggs, 66–67, 70; human reproduction and, 61–62; isogamy and, 66; sizes of, 65–67; sperm, 65–67, 70
sexes, 65–67
sexual reproduction: anisogamy and, 66, 67; bacteria and, 63–65, 70; DNA replication and, 60, 63–64; of eukaryotes, 59–60, 66–67; isogamy, 66; uniqueness of new organisms and, 60; variation and, 64, 65. *See also* human reproduction
sign language, 236–40, 242–43
silicon transistor, 171, *171*, 173
single cells: membranes of, 95; motion of, 81, 82, 84, 90; multicellular organisms and, 95
skeletons: endoskeletons, 104, 105–6; exoskeletons, 104–6, 110

skill formation, 215
Skinner, B. F., 220
skulls: of sauropsids and synapsids, 140; of synapsids, 117; of vertebrates, 117
slugs. *See* sea slugs
snake charmers, Indian, 31
social brain hypothesis, 156–57
social instinct, mammalian, 141, 144, 146, 147, 150, 154
sociality: eusociality, 68. *See also* mammalian social instinct
social life, of humans, 155–57
social media, 227
social structure, eusociality, 68
soil, 100
soma, 69–71
soul, 176–77
Spanish surnames, 61–62
sperm, 65–67, 70
sponge larvae, 84–88, 95. *See also* jellyfish
sponges, 82–88, 95, 96, 118
spores, 101
sporopollenin, 101, 116
squid, 173
squirrels, 151, 152–53
stomata, 100
subjective: mind as, 185; objective and, 178–80
substantia nigra, 212
suffering: of humans, 223–26; meditation and, 226; pleasure, desire and, 224, 226
superorganisms, 69, 71
surnames, 61–62
synapses, 169–71, 174–76
synapsids: body temperature of, 132–33, 147; in Cenozoic era, 139; convergent evolution of, 140; eggs of, 147–48; galloping by, 129, 140; Great Dying and, 120–21, 125–26, 130, 132; heart of, 140; as herbivores, 117; as human ancestors, 132, 147; lungs of, 125–26, 129; maternal care by, 147; of Mesozoic era, 121, 124–26, 129–30, 133, 140, 147; nocturnal bottleneck hypothesis and, 136; nocturnality of, 130–33, 136; offspring of, 147–48; sauropsids compared to, 114, 116–17, 119–21, 125–26, 129, 138, 139–40; skulls of, 117, 140; warm-bloodedness of, 133, 147
synaptic plasticity, 174; cortical columns and, 193–94; memory and, 175; physical, mental and, 179–80

theories, 34
thinking: dopamine and, 227; pattern finding and, 176
Timaeus (Plato), 205
Tolstoy, Leo, 3
top-down information flow: action and, 203–4; computers and, 185; through cortical columns, 199–204; hallucination and, 203–4; human brain and, 183–85, *197*, 207; human perception and, 185, 189, 196
tree of abstractions, 199–202, 204, 241–42, 250
tree of life, ix, 38, 39–42
trees, primate life in, 150–51, 153–54

UCP1 protein, 134–35
unification of world, ix
unity, evolution, diversity and, 36, 250
Urey, Harold, 20

variation, 29–30, 33, 64, 65
vertebrates: arthropods compared to, 99, 105–6; breathing of, 108–9; dominance of, 104; endoskeletons of, 104, 105–6; as herbivores, 117–18; insects and, 98, 104, 105–6, 110–11, 114, 139; jaws of, 117; on land, 98, 104, 105, 118, 141; plants and, 117–18; size of, 110; skulls of, 117. *See also specific vertebrates*

vesicle, 49–51
vesicular transport, 49–51
vision: binocular rivalry, *184*, 184–85, 191; brain and, 207; color sensitivity, 152; depth of, 184; extramission theory of, 205–7; eye positioning and, 152–53; eyes and, 205–7; as fusion of outward and inward fire, 205–6, 207; human perception and, *184*, 184–88, *187*; of humans, 152–53, 184; of mammals, 130–31, 152; neuroscience on, 207; nocturnality and, 130–31, 152; optics of, 205–6; peripheral, 187, *187*; of primates, 152; of squirrels, 152–53
visual cortex, 196–98, 204
vitalism, 22
volcanoes, 120

warm-bloodedness: of animals, 133, 136; body temperature heat threshold of, 136; of mammals, 133–36, 140; of synapsids, 133, 147
water: amphibians and, 114–15, 116; body temperature in, 131; photosynthesis and, 45–46, 102; plants and, 100–103, 116
whales, 129
Wilberforce, Samuel, 26
Woese, Carl, 41–42
wood, digestion of, 125
worms, 96, 98, 109; earthworm, 91–92; food and, 91–94; nemertine, 107–8; Precambrian, 162, 169; roundworms, 110

Yucatan, 137–38

Zen Buddhism, vii, 227